The Rightful Place of Science:

Government & Energy Innovation

The Rightful Place of Science:

Government & Energy Innovation

Edited by

Daniel Sarewitz

Contributors

Miles Brundage
Travis Doom
Jen Fuller
Eric B. Kennedy
Frank N. Laird
Jason O'Leary
Heather M. Ross
Aubrey Wigner

Consortium for Science, Policy & Outcomes
Tempe, AZ and Washington, DC

THE RIGHTFUL PLACE OF SCIENCE:
Government & Energy Innovation

For information on the Rightful Place of Science series,
write to: Consortium for Science, Policy & Outcomes
PO Box 875603, Tempe, AZ 85287-5603

Model citation for this volume:

Sarewitz, D., ed. 2014. *The Rightful Place of Science: Government & Energy Innovation*. Tempe, AZ: Consortium for Science, Policy & Outcomes.

The Rightful Place of Science series explores the complex interactions among science, technology, politics, and the human condition.

Other volumes in this series:

Pielke, Jr., R. 2014. *The Rightful Place of Science: Disasters & Climate Change*. Tempe, AZ: Consortium for Science, Policy & Outcomes.

Alic, John A. 2013. *The Rightful Place of Science: Biofuels*. Tempe, AZ: Consortium for Science, Policy & Outcomes.

Zachary, G. P., ed. 2013. *The Rightful Place of Science: Politics*. Tempe, AZ: Consortium for Science, Policy & Outcomes.

ISBN: 0692297502

ISBN-13: 978-0692297506

LCCN: 2014917540

FIRST EDITION, NOVEMBER 2014

ABOUT THIS BOOK

The Rightful Place of Science: Government & Energy Innovation is part of a larger set of activities centered at Arizona State University's Consortium for Science, Policy & Outcomes (CSPO). This program aims to clarify and provide guidance for energy innovation policy, especially in the context of the need to address global climate change while ensuring access to affordable and abundant energy supplies. Other products from this program include:

- *Innovation Policy for Climate Change: A Report to the Nation* (2009)
- *Four Policy Principles for Energy Innovation and Climate Change: A Synthesis* (2010)
- *The Department of Defense as a Catalyst for Energy Technology Innovation* (2012)
- *The DOD Energy Innovation Atlas* (2013)
- *Our High-Energy Planet* (2014)

These reports can be found at CSPO's website: http://www.cspo.org/areas-of-focus/energy/

Development of *Government & Energy Innovation* was supported by the Bipartisan Policy Center, the Clean Air Task Force, and the Consortium for Science, Policy & Outcomes at Arizona State University.

CONTENTS

1

INTRODUCTION: TWO CENTURIES OF COMPLEX POLICY EVOLUTION

Daniel Sarewitz

The rise of the United States as an industrial power brought with it a continual need for new and better sources of energy and power. Indeed for nearly 200 years the U.S. government has played a central role in advancing energy innovation—a role that has aimed to achieve multiple goals, and has often been politically controversial.

In the first half of the 19th century, steam boilers were an increasingly important source of power for industry, shipping, and rail transport, but this crucial energy technology also had a tendency to blow up, often with devastatingly lethal consequences. By 1830, Congress was funding a private organization, The Franklin Institute in Philadelphia, to perform research on the causes of boiler explosions.[1] Later in that decade the Navy, which was beginning to develop steam-powered ships, hired an academic scientist to lead its own efforts to improve boiler safety. The government also provided funding to determine the energy-generating potential of different coal varieties, information that could aid both government and industry decision-making. By 1849, Congress began to support research on alternative power sources by granting $20,000 to a private citizen (patent examiner Charles Page) to develop an electric motor that could power a railroad locomotive. Such innovation policies were not uniformly popular. They were, for example, opposed by none other than Jefferson Davis, then a senator from Mississippi, who argued that "the government should be the last agent to interfere in these matters."[2]

These earliest forays into government-sponsored energy innovation illustrate three key themes that persist to this day. The first is the great diversity of roles that the government found itself playing, such as researching safety and establishing standards; exploring alternative energy technologies; and providing information on energy supply. The second theme is the importance of the military in advancing energy technology. (Barely a century after the Navy was working to improve coal-fired steam boilers, it was beginning to design submarines with nuclear-reactor-generated steam power). The third is the political disagreement surrounding government energy innovation activities.

But these first efforts also help make clear why energy innovation must command the attention of government. Economic performance and military capability are foundations of political stability, and each is directly linked to the availability and continual improvement of the full range of energy technology applications, from the extraction of energy resources (the first steam engines were designed to pump water from coal mines) to the provision of power for factories, armies, and homes. By the 1960s, environmental protection had become a third principle rationale for government action, initially to address problems of air pollution and the consequences of coal mining and offshore oil drilling, and more recently to address the challenge of climate change. Today, these three fundamental motivations—economic opportunity, national security, and environmental sustainability—provide a continued justification for a strong government role in energy technology innovation. But what, exactly, should that role be?

Answering that is difficult for two reasons. The first is that the process of innovating in the energy domain is incredibly complex. Investments in innovation, whether public or private, are bets on a highly uncertain future, and with uncertainty comes disagreement about what will work and what will not, and about who can and should shoulder the risks that uncertainty creates. From the future of energy prices, to the technical advances that improve competing technologies, to the web of policies and regulations that govern everything from oil drilling to nuclear reactor development to solar energy siting, to unpredictable events (like the Fukushima earthquake, tsunami, and meltdown) that influence consumer preferences and national policy decisions, ongoing and irreducible uncertainty means that there is no such thing as the "right" or "optimal" energy innovation pathway.

The second reason why it's so difficult to define the appropriate role for government in energy innovation, almost eerily foreshadowed by Senator Davis prior to the Civil War, is that political debates about innovation in general, and energy innovation in particular, bring up more fundamental questions about the appropriate limits of governmental action. Today, debates over how to think about the correct role for government in energy innovation tend to map onto not-terribly-useful tropes such as "market failure," "the valley of death," "picking winners and losers," "corporate welfare," "breakthrough innovations," and so on: symbolic concepts that supposedly can help guide decision-making about where government action is necessary, and where things should instead be left to the private sector. The problem with such tropes is that their political value in signaling to voters appropriate ideological credentials is a lot higher than their analytical value in determining how to design good innovation policies.

The cases in this book were developed to contribute to improved public discussions and political debate about how government can best contribute to energy innovation. They aspire to meet science policy scholar Andrew Sterling's challenge to "Keep It Complex"[3]—while also keeping it accessible and comprehensible. Thus, the cases show in detail the many ways in which public investments and an extraordinarily diverse range of policy approaches have been focused on expanding and improving options for harnessing energy for human benefit, even as they also reflect and have been buffeted by changing political winds. The cases illustrate the uncertainties, failures, and ambiguities that accompany efforts to improve energy technologies. But individually and collectively they make clear not only that government has played essential and constructive roles in energy innovation, but also that it must continue to play these roles in support of the overarching goals of economic competitiveness, military security, and environmental protection.

Three especially pertinent themes emerge from the full collection of cases. The first is that the military context continues to provide not only a strong motivation, but also a powerful set of institutional and policy assets, for energy innovation. Indeed, to make clear these assets, our first case, "Department of Defense and Environmental Innovation Pathways," by Aubrey Wigner, is not about energy innovation per se, but about how the Department of Defense organizes and carries out environmental research and development, and why it is able to orchestrate innovation in ways that other areas of government, and even private firms, might want to try to emulate.

This unique institutional capability for strategic innovation policy is on clear display in Travis Doom's study of DOD's long-term efforts to improve the performance and expand the uses of jet engines ("Aeroderivative Gas Turbines"). Here, DOD has over a period of many decades been able to focus the work of private firms to deliver continual innovation that decisively advanced national security, built unparalleled competitive strength for the American aerospace industry, and brought important new technological options into the energy generation mix. The story is one of long-term, mutually beneficial partnership.

Despite such successful stories, energy technologies can become hostage to partisanship. For example, shortly after being elected, President Reagan had the solar panels removed from the White House as one way to differentiate himself from his predecessor. In "The Effects of the Department of Defense on the Photovoltaic Industry," Frank N. Laird shows that even if DOD is not likely to be a major force in directly accelerating solar energy innovation, it may help to tame the politics around energy policy. During more than a decade of desert warfare, widely seen images of solar panels providing energy for American bases help signal to citizens and policymakers alike that solar energy is not just a boutique technology advanced by environmentalists, but also an important contributor to our national capacity to operate effectively on the battlefield. The role of DOD in energy technology is not simply as a strategic innovator, but also as a stabilizing political symbol.

The second theme revealed by these cases is that, when it comes to energy technological innovation, the boundaries between the government and industry are fuzzy, and cooperation and burden-sharing across sectors are essential for both. In the mid twentieth century this idea was strongly embodied in the government's essential role in rural electrification. The Tennessee Valley Authority and Bonneville Power Administration, for example, brought power and the potential for economic and industrial development to large, power-poor regions of the nation. After World War II, government was similarly the prime force behind the development, demonstration, and promotion of civilian nuclear power, while in the past couple of decades, the natural gas revolution in the U.S. has been accomplished in no small part due to government investments in research that supported innovation in gas extraction technologies by the private sector.[4]

This complex mixing of public and private roles is highlighted in two cases exploring solar energy innovation. Miles Brundage's "The Government Role in Developing Solar Thermal Technology" details an array of policy tools and partnerships led by the Department of Energy and also including utility companies and technology start-ups, aimed at developing, demonstrating, and driving down the cost of solar thermal energy. Jason O'Leary's "Federal Investments in Advancing Solar Photovoltaics" traces the government role back to the earliest days of the space race, and shows how policy approaches have evolved and become, along with the technology itself, more complex and sophisticated over time. Initial approaches mostly involving direct research and development contracts to inventors have given way to loan guarantees, tax incentives, and standards-setting in what amounts to a policy portfolio applied toward advancing performance of photovoltaic energy while encouraging the acceleration of private sector innovation and growth of markets.

From these cases a third strong theme thus emerges: the remarkable diversity of policy approaches that government has called upon to foster and accelerate energy innovation. Such diversity is well illustrated in "Electric Vehicle Development and Battery Technology" by Eric B. Kennedy and Heather M. Ross, a story that shows how government has used an array of innovation policies to take advantage of the complementary capacities of the public and private sectors. Here we see government providing not only research and development funding, but also tax credits for consumers, environmental regulations such as fuel economy standards, and even direct support for manufacturing, as part of a suite of policies deployed over several decades to advance innovation and markets. This story includes efforts to forge partnerships between government and the major automobile manufacturers characterized by both political disagreement and difficulties in achieving desired technical advances. Yet there seems little question that without government, advanced battery and electric vehicle innovation in the U.S. would have been much slower, and the nation's competitive status and prospects in this technological domain would be much weaker.

Developing affordable and environmentally beneficial liquid fuels is perhaps the greatest challenge to energy innovation. Jen Fuller examines one innovation pathway toward meeting this challenge in the book's final case, "Biofuels and Cellulosic Ethanol." Here we see government working at all steps of the pathway, from development of crop feedstocks, to supporting the chemistry research necessary for fuel refining, to funding demonstration plants for fuel production, to implementation of appropriate auto fuel standards. This case makes particularly apparent how the variety of policy interventions themselves depend on different institutional arrangements that, given the strong existing technological lock-in around gasoline, would never occur in the absence of public investment.

The cumulative lessons from these cases are neither simple nor prescriptive. They make clear why energy innovation is complicated, why progress is often slow and unpredictable, why mistakes and failures are inescapable—and why the role of government is essential. They add up to a strong case for pursuing multiple technological pathways using multiple innovation policy tools that build ongoing partnerships, and risk-sharing between public and private sectors. They also point to the benefits of long-term persistence and strategic focus, especially as exemplified through the institutional and political attributes of the Department of Defense. The search for the right balance between policy experimentation and strategic focus will remain a challenge for policy makers—and a temptation for politicians. Responsible decision makers in the energy innovation domain will be those who recognize this tension, and can learn how to better manage it by understanding the lessons provided by cases such as the ones presented here.

---

[1] Hunter Dupree, *Science in the Federal Government: A History of Policies and Activities* (Baltimore, MD: Johns Hopkins Press, 1957).

[2] Ibid., p. 49.

[3] Andrew Stirling, "Keep it Complex," *Nature* 468 (23 Dec. 2010): pp. 1029-1031.

[4] Michael Shellenberger, Ted Nordhaus, Alex Trembath, and Jesse Jenkins, *Where the Shale Gas Revolution Came From* (Oakland, CA: The Breakthrough Institute, 2012), available at: http://thebreakthrough.org/index.php/programs/energy-and-climate/where-the-shale-gas-revolution-came-from

2

DEPARTMENT OF DEFENSE AND ENVIRONMENTAL INNOVATION PATHWAYS

Aubrey Wigner

Introduction

The Department of Defense is responsible for numerous environmental remediation sites in the United States. According to a 2010 Government Accountability Office (GAO) report, there are 21,500 sites on active bases, and 11,200 sites on closed bases and areas formally used by the military. In 1990, Congress established the Strategic Environmental Research and Development Program (SERDP), a funding agency administered jointly by the Environmental Protection Agency (EPA), the Department of Energy (DOE), and the Department of Defense (DOD), to address the environmental problems experienced across all of the individual military services and by the DOE. The purpose of SERDP was to identify research that could be usable by both the government and private organizations to address "environmental restoration, waste minimization, hazardous waste substitution, and other environmental concerns, and to share such research, technologies, and other information with such governmental and private organizations."[1] In 1995, the Office of the Secretary of Defense added a sister organization, the Environmental Science and Technology Certification Program (ESTCP), which focuses on demonstration and large-scale testing of environmental remediation, and more recently, sustainable energy, technologies.[2] SERDP and ESTCP are co-located and operate under a single management structure, but they serve very different innovation purposes within the DOD. In contrast to SERDP's multi-agency administrative structure, ESTCP is purely a DOD-administered organization, albeit with strong ties to SERDP and regular contact with the EPA and DOE. While SERDP funds mostly basic research, ESTCP focuses on moving early stage technologies across the "valley of death," where innovations often fail due to a lack of financing and large-scale demonstration in real world conditions. The organizations are funded separately, but at similar levels, with SERDP receiving $66 million and ESTCP $62 million in FY 2012. It is worth noting that ESTCP experienced a $22 million funding increase in 2012 to expand into large-scale energy testing in addition to its previous environmental technology testing. Prior to 2012, ESTCP had been funded at roughly one-third to one-half of SERDP's budget. To date, SERDP and ESTCP have funded nearly 1000 completed projects and are currently pursuing another 500 projects.

In the arena of groundwater remediation, the focus of this study, SERDP and ESTCP have been particularly successful in researching, developing, and demonstrating innovative technologies. Since 1990, SERDP has completed 119 groundwater contamination studies. ESTCP has completed 99 since 1995. To substantiate claims of success in the innovation sector, it is important to have metrics for success and failure. For late stage technological demonstrations, this study uses the metric of deployment rate, that is, the percentage of projects that reach real world field or pilot testing, for comparison between ESTCP and other innovation organizations. SERDP on the other hand focuses almost entirely on the production of early stage research and the dissemination of research findings. In this case, their success is shown by the number of publications, tools for analysis produced, and stakeholders networked for the purpose of greening the DOD. In regards to groundwater remediation, ESTCP has achieved a deployment rate of 79%, which is quite impressive when compared to the historical rates of success seen by another government environmental remediation program, the DOE's Office of Science and Technology, which managed between 28% and 40% deployment from 1989 to 1997.[3] ESTCP's record also compares favorably to another environmental cleanup program, the EPA Superfund program, which managed a 59% deployment rate as of 1998,[4] and even to the highly regarded Advanced Technology Program at the National Institute of Standards and Technology, which saw 64% of its funded projects reach commercial prototypes.[5] SERDP has published groundwater remediation research and technical documents covering a wide range of technologies, including all of the 15 methods pointed to by a 2005 GAO study as the full range of currently available technologies for groundwater remediation.[6]

Given the complexities of environmental problem solving, and broader challenges of translating knowledge and innovation into desired outcomes, SERDP and ESTCP potentially offer important organizational and policy lessons about the management of the innovation process. SERDP and ESTCP share a strong reputation among academics, environmental consultants, and the military for excellent management and performance across all domains of innovation, from fundamental research to demonstration and deployment. According to interview data and document analysis, the DOD, the EPA, the DOE, the Federal

Remediation Technologies Roundtable (FRTR), and a wide range of industry consultants in the remediation business all make regular use of SERDP and ESTCP scientific reports, cost reports, and guidance documents.

This paper investigates SERDP and ESTCP's efforts in groundwater remediation in order to better understand and describe the programs' approach to problem-solving. Groundwater remediation was chosen for several reasons. First, it has been one of the longest running areas of study pursued by SERDP and ESTCP, going back to the founding of both programs. With 91 current projects and 213 completed projects, there is abundant data for longitudinal analysis as well. By analyzing data trending back to the founding of SERDP in 1990 and ESTCP in 1995, a clear picture emerges of the organization's methods of innovation and the success or failure of their projects. Finally, this extensive experience with water remediation projects has played a large role in establishing, and continues to influence, SERDP and ESTCP's understanding and management of innovation.

Since the formation of SERDP and ESTCP, the methods used for remediation of contaminated sites throughout the DOD complex have changed drastically. As the 2005 GAO report highlights, the DOD has transitioned from decades of using expensive and time-consuming pump-and-treat technologies to a broader strategy of developing, using, and testing all available technologies for groundwater remediation.[7] SERDP and ESTCP have aided the DOD by providing basic research, field trials, cost analyses, planning and deployment guidelines, and the R&D to industry networking necessary for all 15 of the remediation technologies highlighted by GAO. This provides a stark contrast to a previous GAO study in 1994, which identified barriers such as limited cooperation between the DOD and other agencies and inconsistent research funding methods as major limits to the abilities of the DOD to conduct environmental cleanup.[8] While it is unlikely that SERDP and ESTCP are solely responsible for this transformation in DOD remediation efforts they have certainly played a crucial role by advancing the state of the art in environmental remediation technologies by researching, developing, and disseminating guidance documents for the use of technological alternatives to classic pump-and-treat systems such as in-situ bioremediation, natural monitored attenuation, and the focusing on the identification and remediation of dense non-aqueous phase liquids (DNAPLs). Since 1990, SERDP- and ESTCP-funded projects have tested dozens of methods for contaminant detection, published guidance documents, developed software for the qualification of contaminated sites as well as models for cleanup, and deployed hundreds of real world tests from field demonstrations to pilots and full scale clean-up projects across dozens of military sites in the United States.

This study aims to determine what innovation pathway has made SERDP and ESTCP so successful at providing quality environmental science to the defense community and the associated private organizations which service it. By studying the internal workings of a successful government research and development organization, I hope to provide organizational principles that may be relevant for, and applied to, other areas of innovation, especially those that demand a strong government role, such as energy innovation.

Motivation for Study

Currently, the United States faces many challenges that seem to require technological solutions. Problems ranging from climate change and energy infrastructure to sustainable natural resources management and affordable medical care all attract significant public investments in research and development. Such problems are similar in that they are "wicked": highly complex, strongly context-dependent, politically divisive, and never "solved" in any conventional sense of the term. Determining the innovation pathways that are most likely to contribute to effective management of such problems would have great economic, political, social, and environmental benefits. For the purposes of this paper, an innovation pathway will be defined as the methodology used by organizations involved in researching, developing, and disseminating information about technologies. SERDP and ESTCP have been highly effective in the specific area of environmental remediation, and the lessons from this record may offer more general insights into successful innovation practices. Unlike many organizations that focus on individual aspects of the innovation process or on broad scientific fields, SERDP and ESTCP have developed what might be termed an "ecological" approach to R&D, one that is not merely about the production of knowledge, but also its testing and dissemination, as well as a mission to foster the relationships that allow for this knowledge to be used effectively in real-world situations. For SERDP and ESTCP, the goal is to produce knowledge for explicit use in solving environmental issues in the defense sector. Although these programs lie at the intersection of three distinct U.S. government mission agencies, the Department of Defense, the Department of Energy, and the Environmental Protection Agency, they have managed to effectively navigate this organizational diversity and indeed benefit from it by exploiting the broad range of regulatory expertise and scientific knowledge that it offers, and by creating cross-agency test sites and reports. The result is a unique approach to innovation for solving the specific, large-scale problems that face the DOD in the realm of environmental R&D.

Background of SERDP and ESTCP

In an address to Congress in 1990, Sen. Samuel Nunn (D-Ga) put forth the idea of environmental security:

> *There is also a new and different threat to our national security emerging – the destruction of our environment. The defense establishment has a clear stake in countering this growing threat. I believe that one of our key national security objectives must be to reverse the accelerating pace of environmental destruction around the globe.*[9]

SERDP was first envisioned and introduced to the Senate via the Strategic Research Program Act of 1990. This act, sponsored by Senator Nunn, was a mandate for the Office of the Secretary of Defense to form an organization dedicated to researching and providing solutions for the Department of Defense's environmental problems. SERDP was the result. While the Strategic Research Program Act of 1990 stalled in the Senate, the language used in it was adopted almost verbatim for the National Defense Authorization Act for Fiscal Year 1991 (PL 101-510), which formalized SERDP as an organization within the DOD.

SERDP funding comes entirely from DOD, with high level governance and oversight provided by the SERDP Council, which is composed of representatives from the Office of the Secretary of Defense and each of the individual services, as well as from DOE and EPA. The SERDP Council not only acts as the sole funding authority in SERDP, but also seeks to "maximize exchange of information and to minimize duplication of environmental research activities through close coordination with the military departments and Defense agencies, the DOE, the EPA, and other departments and agencies of the federal, state, and local governments."[10] Under this arrangement, SERDP can benefit from the regulatory advice and scientific expertise of the EPA and the DOE available through all stages of the innovation process. An overview of SERDP's organizational structure is shown below.

Figure 1. SERDP Organizational Chart

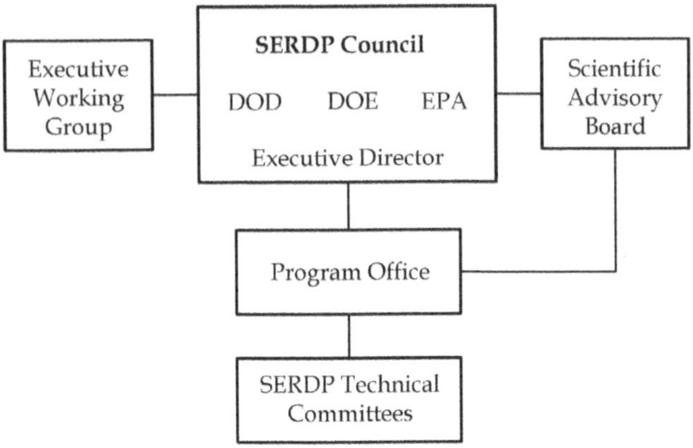

Source: Adapted from www.serdp.org/About-SERDP-and-ESTCP/About-SERDP.

The members of the SERDP Scientific Advisory Board are jointly appointed by the Secretary of Defense and the Secretary of Energy, with advice from the EPA, and are drawn from academia, government, and industry. For example, currently the science advisory board is composed of six academics, three industry experts, a representative from the National Oceanic and Atmospheric Administration (NOAA), a member of the White House Office of Science and Technology Policy, and chaired by a state official, the director of the Nebraska Department of Environmental Quality. The Scientific Advisory Board is considered a formal Federal Advisory Committee; SERDP Council chooses the Board's members under these guidelines:

> *To ensure that SERDP objectives are consistent with the Administration's goals, two members of the SAB are mandated in the statute – the Science Advisor to the President, and the Administrator of the National Oceanic and Atmospheric Administration, or their designees. Also, to ensure that regional and global environmental issues are appropriately addressed in SERDP, at least one member should represent the interests of State governments and one member should represent environmental public interest groups.*[11]

Technical Committees are formed from experts in each of the five technology programs SERDP pursues. Much like the members of the Scientific Advisory Board, these experts are drawn from a wide range of government, industry, and academic backgrounds. However, Technical Committees have no specific requirements for representing any given organization or sector. They are formed through a network of experts in academia and industry and are brought together to act as expert advisors on the areas of priority research as determined by the SERDP Council. The representatives of the services and agencies that serve on the SERDP Council select the individual committee members. Each year the Council, in collaboration with the Executive Working Group (composed of technical experts from each of the services and organizations represented on the Council), decides which areas are priorities for DOD research. From there, with the aid of the Scientific Advisory Board and SERDP Technical Committees, yearly Statements of Need (SONs) are issued. The SONs define what project areas SERDP will be funding for the year. In the case of groundwater remediation, each SON gives background for the problem, the expected payoff that should result from the research, and, for some SONs, the expected amount of funding SERDP will provide. Program applicants are welcome from both industry and academia. Examples of SONs can be seen at SERDP's website under the funding section.[12] This view of their innovation methodology is illustrated below, using a flowchart from a 2006 presentation on SERDP and ESTCP's investment strategy, given by Dr. Andrea Leeson, the program manager for environmental restoration.[13]

Figure 2. Service Coordination Process

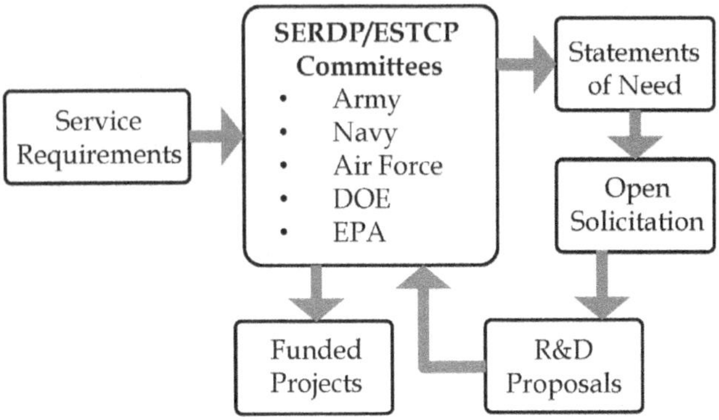

Source: Adapted from Andrea Leeson, "Investment Strategy in SERDP and ESTCP," presentation, May 2006.

As shown in the Service Coordination Process Flowchart, the individual military services are engaged in the SERDP innovation process in several ways. First and foremost, each service has a representative on both the SERDP Council and the Executive Working Group. Second, the services themselves produce yearly requirements for remediation goals, technologies that could use environmentally friendly alternatives, energy consumption/production goals for facilities, and so on. These requirements are taken into account by SERDP and ESTCP when producing each year's Statements of Need. While the above flowchart shows the Services' requirements as being a one-way input into the coordination process, it would be more accurate to view this as a two-way information exchange. Service requirements go out to a wide range of innovators in the DOD including individual service laboratories and external consultants. SERDP, according to interviewees, attempts to maintain two-way communications with other military labs in order to avoid the duplication of research work and to collaborate on possible projects that are of interest to multiple parties.

The project proposal and funding selection processes are described in greater detail in the Findings section of this paper. At this point, it suffices to say that SONs are released to the public, proposals are returned from academic and industry partners, and the final funding decisions are made by the SERDP Council.

The dissemination of SERDP- and ESTC-funded research takes place in several ways. SERDP and ESTCP hold a yearly symposium where researchers and industry partners are able to see the results of funded projects as well as pitch other solutions to DOD environmental problems. In addition to this yearly symposium, workshops are held for specific problems in which, once again, researchers and members of industry are invited to collaborate. Both the symposiums and workshops will be discussed in greater detail in the analysis portion of this paper. All the results of SERDP and ESTCP research are freely available online and in published volumes covering specific research areas. Furthermore any tools and findings developed through funded research are also freely available to aid in planning environmental cleanup projects or to characterize sites. Under the Patent and Trademark (Bayh-Dole) Act Amendments of 1980, researchers with federal funding may maintain intellectual property rights to their discoveries and studies, though the government can freely use them without patent restriction. However, the vast majority of SERDP and ESTCP study data is available with no copyright or patent restrictions, though remediation companies would be well advised to check for applicable patents and licensing issues before adopting a SERDP-developed technology as there are cases where individual companies or researchers may be using patented methods or materials.

Methods

I used two primary methods in the analysis of SERDP and ESTCP's groundwater remediation program. The first was a basic document analysis, drawing on publicly available information on the organizations, including organization websites, government memos and laws governing them, presentations by the organizations, presentations delivered at workshops and the yearly symposium, published studies and reports, and, finally, assessment documents from outside organizations, especially the Government Accountability Office. In total, I analyzed over 100 reports, from yearly budget reporting to GAO studies. The SERDP and ESTCP executive director and environmental program manager provided additional background. Second, I conducted in-depth interviews with several key informants who have worked closely with the two organizations, either as primary investigators on research projects, as members of technical boards where statements of needs were developed for the yearly projects, or both. (Two interviewees had served on technical boards and had also been primary investigators.) A total of five interviews were conducted. The interviews were conducted in an open-ended format. Questions focused on determining how SERDP and ESTCP were different from other funding organizations the interviewees had worked with and what the internal processes of SERDP and ESTCP were with regards to choosing projects, networking with the broader remediation community, and reporting requirements for funded projects. The interviews were analyzed qualitatively to identify crosscutting ideas, themes, and observations that inform, deepen, and modify the portrayal of the SERDP and ESTPC innovation process derived from publicly available sources. Additional document analysis was then conducted to see how these observations could help to form a more in-depth picture of the internal workings of SERDP and ESTCP.

Findings

The five interviews strongly converged on a small set of common attributes and activities that characterized the SERDP/ESTCP innovation pathways. These commonalities were mentioned by all interviewees, and form the framework of my analysis:

- In-House Innovation Qualities
  - Mission-focused research goals
  - Straightforward application process
  - Stringent reporting requirements for research progress
  - Siting support for technology testing
- Knowledge and Technology Transfer Expertise
  - The production of real-world cost estimates and contaminant categorization tools
  - Networking support through the yearly symposium and regular technical workshops

The four items grouped together under "In-House Innovation" are funding and research support activities and organizational qualities at which SERDP and ESTCP appear to excel. The two items listed under "Knowledge and Technology Transfer" focus on what SERDP and ESTCP do to disseminate completed research and organize a community for the remediation of DOD sites. Interviewees also discussed how SERDP and ESTCP compared to other, similar organizations, such as the DOE's Environmental Management Program's Office of Science and Technology (OST), which the GAO has previously compared to ESTCP and which I will discuss in more detail below.

To more fully flesh out the general description of SERDP and ESTCP that emerges from the document analysis and interviews, I have also developed three brief case studies that both exemplify and illustrate in detail how SERDP and ESTCP function and how they differ from other organizations pursuing similar innovation goals. The first case discusses a specific contaminant, perchlorates, the second discusses a successful series of SERDP projects, the National Environmental Technology Test Sites, and the final case discusses DOE's Office of Science and Technology, which was founded at about the same time as SERDP and ESTCP, and had a similar mission.

In-House Innovation

The SERDP and ESTCP mission is R&D to address the DOD's environmental problems, though the two organizations approach this mission with slightly different focuses:

> *SERDP focuses on cross-Service requirements and pursues solutions to the Department's environmental challenges. The development and application of innovative environmental technologies will reduce the costs, environmental risks, and time required to resolve environmental problems while, at the same time, enhancing and sustaining military readiness.*[14]

> *The Program (ETCP) was established in 1995 to promote the transfer of innovative technologies that have successfully established proof of concept to field or production use.*

> *ESTCP demonstrations collect cost and performance data to overcome the barriers to employ an innovative technology because of concerns regarding technical or programmatic risk, the so-called "Valley of Death."*[15]

SERDP maintains strong links with each of the military services and as a result is further able to narrow down the scope of specific problems. SERDP and ESTCP have five specific focus areas: Energy and Water, Environmental Restoration, Munitions Response, Climate Conservation and Climate Change, and Weapon Systems and Platforms. Contaminated Groundwater, the focus area of this study, falls under the broader envelope of Environmental Restoration. By focusing on these program areas as they pertain to specific DOD sites SERDP/ESTCP managers, with feedback and input from specific DOD site managers, are able to clearly define problem areas to produce SONs for specific solutions. For example, one of the main areas of interest for groundwater remediation is the issue of Dense Non-Aqueous Phase Liquids (DNAPLs), contaminants which are denser than water and which pool in aquifers and are thus difficult to identify and remediate. One major source of DNAPL contamination is chlorinated solvents, which the military used extensively after World War II. By discussing contaminants of concern with the individual services, SERDP was able to identify the need for, and focus research funds towards, the remediation of chlorinated solvents early on and thus produce a body of literature and tools to aid in remediating these chemicals.[16] One interviewee identified this early-focusing aspect of the SERDP/ESTCP innovation pathway as primary to the program's success. He had previously served on the Science Advisory Board and participated in its "research needs" meetings to determine the technologies or contaminants that would be the focus of SERDP SONs. Effective mission focus is aided by the role of EPA on the SERDP Council, which helps to create a clear communication pathway from EPA regulation on hazardous materials, through DOD analysis of which sites contain these materials, to the need for researching techniques that will aid in remediation. EPA representation on the SERDP Council allows for clear lines of communication regarding regulation and compliance with environmental standards. In contrast, the DOE representative's part seems more related to the commonality of environmental problems on DOD and DOE sites. Furthermore, since the end goal of any research by SERDP/ESTCP is to actually remediate DOD sites, expected cost for large scale deployment ends up being a primary focus of any SON, particularly those released by ESTCP. This is not to say that SERDP and ESTCP are uninterested in basic research regarding contamination, rather, since the end goal is cleanup, projects that focus on developing low cost cleanup solutions are preferentially funded. The same interviewee pointed out that, while DoE and EPA had also both pursued quality research in the remediation area, they tended to focus on laboratory science rather than field science. In the case of groundwater remediation, for example, this might have led to an improved categorization of hazardous materials, but very little research that would be applicable for actual large-scale cleanup at the level required by the DOD. For example, in the case of perchlorate (see text box), EPA has played a role in the regulatory arena and in defining contamination testing and concentration limits, but has not made significant strides in actual cleanup technologies.

Box 1. Perchlorate Remediation

A primary example of SERP's proactive and collaborative approach to environmental cleanup can be seen in its research on perchlorate contamination in groundwater. Perchlorate is a compound that is used primarily in rocket fuel and explosives. Starting in 1998, perchlorate was added as a candidate for regulation by EPA under the Safe Drinking Water Act (SDWA). At the time it was added however, little was known about perchlorate's behavior in groundwater—including whether it was entirely manmade or could also occur naturally, and what levels of concentration would be unsafe for drinking. That same year, SERDP began investigating methods for detection and cleanup of perchlorate.[17] While perchlorate is still being considered for federal regulation by the EPA[18], SERDP and ESTCP have performed numerous studies on how to detect perchlorate, what its effects are on plants and wildlife, and how to remediate it using a variety of in-situ techniques. Part of the driving force for SERDP's investigation into perchlorate remediation has been state-level regulation of the contaminant. Since the DOD has sites in every state, any individual state's environmental legislation can help shape SERDP and ESTCP research goals. In particular, California's Perchlorate Contamination Prevention Act of 2003 played a strong role in making perchlorate identification and remediation a priority for the DOD due to the number of active and former military sites in the state.[19] Massachusetts and at least ten other states have also enacted perchlorate regulations.[20]

SERDP and ESTCP have published 117 final reports on perchlorate-related studies, 26 cost and performance reports, and 10 guidance documents, which are in use by the DOD for perchlorate remediation. All of these reports and guidance documents have been produced despite the lack of federal regulatory standards for perchlorate. SERDP studies have also helped to inform EPA's standard-setting process by developing new technologies for the detection and monitoring of perchlorate concentrations. According to a 2010 GAO study, the DOD reported that 407 installations had sampled for perchlorate between 1997 and 2009. An additional 361 installations reported not testing for perchlorate due to a historical lack of perchlorate-producing compounds being used, stored, or produced at their locations. Of the 407 sites that tested for perchlorate, 53 had contamination levels above 15 parts per billion, an EPA suggested cutoff-point for groundwater safety. During that same period, two sites reported completion of their remediation goals, 13 sites were in the remediation process, 31 sites were assessing their remediation options or engaging in ongoing monitoring, and 7 sites took no additional action after consulting with local regulators.[21] ESTCP is involved in projects ranging from field-testing to large scale remediation at a variety of both DOD and DOE sites, including, but not limited to, some of the sites listed in the 2010 GAO study. Work at SERDP and ESTCP for perchlorate remediation is an ongoing process with 4 currently active SERDP studies and 15 ESTCP demonstrations.

The application process for SERDP/ESTCP-funded projects is particularly straightforward. After a SON is released, applicants can review the stated needs and then forward their project proposal to HydroGeoLogic, a firm hired by SERDP to review proposals and do the background research to insure only high quality proposals are forwarded to the Science Advisory Board, and eventually the Executive Director and his staff, for final approval. While the SONs, as mentioned, put a high priority on estimated final deployment costs, they do not specify budgetary requirements for program applicants. They instead simply specify what types of research they are interested in, and allow the individual applicants to propose an appropriate budget. SERDP and ESTCP projects typically range from about $85,000 to $2.5 million per year.

In addition to the core project application process of responding to a SON, SERDP also has the SERDP Exploratory Development (SEED) program to support one-year, small scale proof-of-concept-style projects funded for no more than $150,000.[22] SEED SONs are released at the same time as SERDP core SONs and focus on a few specific areas of interest that the SERDP would like to investigate for that year. For example, for fiscal year 2014, SERDP released two SEED SONs, one for underwater munitions detection and remediation and a second for environmentally sustainable binders for explosives. Successful SEED projects are then considered for full SERDP project funding, though further funding is not guaranteed and must be applied for via the normal core solicitation process. This flexible funding structure allows for a wide range of applicants and projects to be integrated into the SERDP innovation pathway. ESTCP, due to its focus on larger scale field and pilot testing, has no equivalent to the SEED program.

SERDP and ESTCP streamline the application process and reduce the administrative burdens on applicants and on its review process through the use of pre-proposals, proposal presentations, and a rapid turnaround of proposal evaluations. The initial application is a five-page project description with one annual submission date in January. Within one month, pre-proposals are evaluated by HydroGeoLogic and requests for full proposals are sent out. The final proposal is due one month later. Full proposals are reviewed by HydroGeoLogic as well, and applicants who receive favorable reviews present their project orally to the SERDP Scientific Advisory Board six months later. Final project funding decisions are made by the SERDP Council and Executive Director. Project awards are made early in the next year. ESTCP follows a similar process and schedule, though staggered three months later than SERDP's schedule to allow for coordination between the programs. The time from submission of first pre-proposal to project inception is typically about 14 months. According to one interviewee, earlier in the program's history all SERDP proposals and pre-proposals were reviewed by the SERDP program managers and executive director for approval. However, as the organization, and thus the number of submissions grew, an outside source was sought for evaluating the proposals. HydroGeoLogic, an environmental consulting firm formed in 1987, had staff who had been involved with SERDP since its inception both as experts who had served on technical committees and advisors to the executive director, as well as staff who had successfully completed projects for SERDP. As a result of these previous partnerships, HydroGeoLogic was chosen as a trusted source for project evaluation and continues to maintain close contact with SERDP and ESTCP for both proposal reviewing and contracting for a wide range of research activities.

The award process is worth further discussion as it is integral to understanding the strategic approach that SERDP and ESTCP take to achieving their mission. Interviewees repeatedly mentioned that SERDP takes a fairly holistic view on funding projects. That is to say, they focus not merely on scientific merit or covering a broad selection of different methodologies, but also on how to combine related proposals into a collaborative group in order to help a technology grow. By looking at each project as part of a collective effort to solve a mission goal, they are able to better guide the collaborative process later on. Furthermore, SERDP in particular tries to achieve a balance in its funded projects among high-risk conceptual work, basic research to provide fundamental knowledge for problem-solving, and safer incremental advances. ESTCP however functions much more on the demonstration end of projects, thus any ESTCP project already needs to have firm footing and demonstrated success on a laboratory scale. Some, though not the majority, of ESTCP projects have their roots in SERDP studies, and those that do are often not performed by the same researchers. For example, Generation of Biodegradation-Sorption Barriers for Munitions Constituents (project designation ER-201123), a research study on enhanced biodegredation of contaminants at live explosives testing ranges is based heavily on both a previous SERDP study as well as a previous ESTCP demonstration.[23] The project aims to determine the efficacy of enhanced biodegredation methods for decreasing contaminant build-up on active ranges. Neither the previous SERDP study nor the ESTCP demonstrations were done by the same PI or even the same institution. This example offers an excellent view into the SERDP and ESTCP approach to innovation. By emphasizing communication among its funded community, open access to all results, and a strategic approach to funding decisions, SERDP and ESTCP are able to bring in new investigators at various stages of a project, to continue and expand on the work of the studies done before, even if the new researchers had no direct links to the original study. Finally, while SERDP receives Congressional Adds, or "earmarks," the earmarked portion of SERDP's budget is less than 1%,[24] and as one interviewee put it, "earmarked project undergo the same rigorous screening for scientific merit as any other project, they just end up going to specific states."

SERDP and ESTCP provide what amounts to project management services for the researchers by engaging them continually in the research process. In keeping with their mission-focused nature, SERDP and ESTCP require more frequent output and financial reports, which improves project outcomes through more frequent feedback from program managers, potential mid-project funding adjustments, and more accountability. SERDP requires monthly financial reports as well as quarterly progress reports, both of which can be filed online via SERDP's website and are, according to interviewed PIs, straightforward. Finally, two types of briefings are also required, one is a yearly In-Progress Review meeting, and the second is an End-Of-Project Briefing, both of which are presented to the program area managers, e.g., the Environmental Restoration program manager, the Energy and Water program manager, etc. All interviewees agreed that the reporting process was not particularly onerous but more stringent than other funding organizations they had dealt with. Overall, the interviewees on the consulting side felt SERDP's reporting mechanisms were highly effective in helping to guide the projects and maintain communications between SERDP managers and the primary investigators. While nobody in the survey group voiced complaint over the reporting process, one interviewee did say that he thought the reporting might be somewhat outside of what academics normally dealt with. He regarded this as a good thing and further reinforced the idea that regular reporting was something that allowed not only for SERDP to track project progress, but also to provide regular feedback from the investigator to SERDP about project needs. He suggested that academics are notoriously poor at project management, and that SERDP and ESTCP helped the research process by providing a strong managerial framework. In this light, an important additional aspect of SERDP's project administration approach is flexibility with regard to funding for ongoing projects, so that adjustments to funding can be made during a project's duration based on emerging opportunities, needs, and problems. The combination of regular reporting, feedbacks between program management and performance, and flexible funding also ensures a high degree of accountability.

For researchers demonstrating a viable remediation technology, finding an appropriate site for live testing can often be difficult. If the process of finding a test site is overly onerous, it can result in project delays, lack of real world examples for end users to base their expectations on, and uncertainty about actual performance metrics for a given technology. Thus, siting support is a vital part of the process in demonstration-scale testing of groundwater and, more broadly, environmental remediation processes. Early on, SERDP recognized the need for testing spaces. As a result, the National Environmental Technology Test Sites (NETTS) were formed. These four national test sites have hosted numerous projects from ESTCP as well as the DOE, EPA, and industry partners. The sites and their background are described in greater detail in the text box below. One of ESTCP's primary roles when identifying projects worth funding is to network the project managers with an appropriate DOD facility or test facility for the demonstration. As a 2005 Government Accountability Office (GAO) reports states, "[ESTCP] provides a site on a DOD installation where the researcher can field-test the technology and document the technology's cost, performance, and reliability."[25] These demonstrations on actual DOD sites are then used to define baseline requirements for environmental cleanup costs and certify which technologies are preferable.

ESTCP's role in siting support for demonstration projects distinguishes it from previous similar programs, such as the Department of Energy's now defunct Office of Science and Technology (OST), which was roundly criticized by the GAO and in Senate hearings for its ineffectual use of funding for developing environmental cleanup technologies. A 1998 report by the GAO highlights not only the failings of the OST program after its first eight years of operation, with regard to project management, end-user engagement, performance tracking, and technology deployment, but also how well the then-three-year-old ESTCP program was doing.[26] One final note is that OST was specifically criticized for its lack of engagement with end users. In contrast ESTCP and SERDP have, according to interviewees, the Executive Director, and my own observations at a yearly SERDP/ESTCP symposium, made it a point to involve end users, regulators, and the cleanup management personnel at contaminated DOD sites throughout the R&D process.

Box 2. National Environmental Technology Test Site

One of the hurdles to testing new remediation technologies is the lack of accessible sites for mid- to large-scale testing. In order to fully test a technology, sites are needed which contain the contaminants of interest, cover a reasonably large area, and are under the same continuous management for many years. Military, and formerly military, installations provide all of these attributes. Furthermore, military sites also tend to possess excellent documentation about which contaminants are present. One of the earliest SERDP projects, The National Environmental Technology Test Site (NETTS), started in 1993 and currently is composed of four sites on military and former military installations: Naval Base Ventura County (NBVC), Port Hueneme, CA; the former Wurtsmith Air Force Base (AFB), MI; Dover AFB, Delaware; and McClellan AFB, CA. Each site has historical characterization for the contaminants present, monitoring stations, and a variety of test sites. These sites are sponsored and operated by SERDP, but designed to be accessible by the military services, DOE, EPA, and academic and industry partners for technology testing and validation purposes. These sites were also intended to promote technology transfer by allowing a variety of actors to use them for research. Between 1993 and 2007 twenty-five research studies took place at the four sites, carried out by the military services, EPA, DOE, and independent university and industrial researchers. The findings in these studies helped form the groundwork for DOD technical standards and recommended cleanup methodologies. For example, one study on the use of direct-push wells for contamination monitoring instead of conventional wells, resulted in two American Society of Testing and Materials (ASTM) standards and an Interstate Technology & Regulatory Council (ITRC) Technical Regulatory Guide publication. Direct push wells involve pushing or hammering in pipes for testing wells rather than traditional, and expensive, drilling methods. The estimated savings for the DOD from this research ranges from $12 million to $80 million per year in cleanup monitoring costs.[27] Other studies performed at the NETTS sites have given the DOD, EPA, and DOE greater insight into bioremediation of contaminants, air sparging (a remediation process which injects air into groundwater to enhance natural degradation of contaminants), and reactive barrier technologies, all of which are currently in use in a variety of cleanup sites.

Knowledge and Technology Transfer

The ESTCP program helps to certify which technologies are useful and acceptable for DOD site cleanup; informs the services, DOE, and EPA about technological suitability for site cleanup; and produces cost estimates for technology deployment. The cost estimates are developed through large-scale demonstration site studies funded by ESTCP. The cost estimates, and the methodology used to reach them, are all available on SERDP's website and are often accompanied by guidance documents and in some cases by modeling tools as well. These reports aid DOD site managers in determining how to characterize and estimate cost for a cleaning up particular sites.

These cost and performance documents help site managers, whether military or civilian, understand the costs and expected rates of remediation that each technology might provide. It is important to note that the vast majority of environmental remediation research is at least partly the result of federal funding. Without the cost and performance results available for government test sites, there would be little information on the real world performance of advanced remediation technologies. SERDP's and ESTCP's results, along with the research done by other governmental organizations, are published in documents such as the *Guide to Documenting and Managing Cost and Performance Information for Remediation Projects,* produced by the Federal Remediation Technologies Roundtable (FRTR),[28] as well as for individual service internal documents and websites such as the Air Force Center for Engineering and the Environment. The FRTR offers a particularly valuable example of how SERDP and ESTCP findings are disseminated to a broader governmental audience. FRTR is a partnership among the DOD, DOE, Department of the Interior, NASA, the Interstate Technology and Regulatory Council (a public-private partnership focused on fostering the use of innovative environmental remediation technologies) and a variety of other government research labs and centers, for a total of 51 departments, agencies, and laboratories. FRTR convenes a semi-annual meeting to discuss priority cleanup issues, experiences and research results, and to form working groups to solve problems collaboratively. FRTR also publishes guidance for site managers. While the FRTR does not have any sort of regulatory power for dictating which remediation methods and technologies are used, it does act as a trusted clearinghouse for remediation information. For example, the FRTR Technology Screening Matrix presents a visual method for site managers to determine which of 64 available groundwater cleanup technologies would be most appropriate for their site.[29] Through these and other publications, ESTCP's funded projects, in essence, provide a foundation on which DOD and other federal environmental cleanup projects are judged. Additionally, SERDP and ESTCP's close partnership with organizations such as the FRTR and ITRC ensures that their findings are widely disseminated for application not only to government but also private sector remediation projects. For example, as mentioned above in the Perchlorate Case Study, ESTCP produced 10 Guidance Reports and 26 Cost and Performance Reports for perchlorate remediation technologies. These reports supply real-world estimates for the cost of remediation, and they suggest methods for determining a remediation plan, identifying contaminants, and making decisions on which technologies are most effective in a given scenario. Without the cleanup estimates provided by ESTCP, DOD would have difficulty comparing costs for various technologies both in terms of dollars as well as the time needed for complete remediation. As discussed in the comparison case study between ESTCP and the DOE's OST (see text box below), regular deployment of innovative technologies, and good reporting of the results of these deployments, is vital for determining which technologies to use on any given remediation site.

SERDP and ESTCP also offer several vehicles for encouraging collaboration between different research teams, and between academia and industry, the most important of which are their yearly symposium and regular workshops. The yearly symposium offers an opportunity for members from the military, academia, and industry to meet, network, and learn about and discuss the latest research being done by SERDP and ESTCP. In contrast, the workshops, which are organized by Program Area Managers, are designed to bring small groups of specialists together to define and prioritize the research needs surrounding a given specific environmental concern. Both the workshops and symposium will be described in greater detail below. Moreover, as discussed, an additional, ad hoc mechanism for promoting collaboration is SERDP's review process, which occasionally leads to the combining of related proposals from different institutions or sectors, and in other cases makes funding contingent upon demonstrated collaboration among multiple researchers working on similar technologies. The annual SERDP symposium brings together over a thousand researchers, consultants, industry experts, and administrators from academia, industry, and government to present research results and facilitate problem-oriented networking, In doing so, the symposium facilitates communication and collaboration across both sectoral boundaries and different stages of the innovation process, and thus aids the end goal of promoting environmental technology innovation and technology transfer. Technical sessions report on the latest research results and technological developments coming out of SERDP/ESTCP-sponsored products. Short courses are held to train environmental cleanup managers in SERDP-developed technologies, and to educate researchers about the current state of practice. Training is also offered in cost estimation and site categorization methods. A poster session helps further spread awareness about current research sponsored by SERDP, ESTCP and industrial partners. The symposia are thus designed with an appreciation for the complexity of knowledge and technology transfer, with attention toward diverse aspects of, and participants in, the process, including not just transmission of research results but also training, brainstorming and networking. One interviewee pointed out that the symposia further act as an incentive for rapid project completion by fostering a competitive research environment through the sharing of knowledge. The interviewee suggested that this open forum for discussion and presentation forced academic researchers in particular to realize that their work needed to be completed on a reasonable schedule in order to maintain competitiveness for future research funding. From my own observations during a groundwater remediation breakout session, both industrial consultants as well as academics found it useful to discuss current technologies and research in a small group environment focused on topics of interest. These breakout sessions acted as stimuli for technical discussion as well as an opportunity for networking among the government, academic, and industrial interests that must effectively coordinate their activities, and learn from one another, if innovation is to be successful. During the session, one academic expressed frustration about how difficult it was to find an industry partner and a test site for a project. Another member of the breakout session immediately responded that he was looking for research for his company to test, and the SERDP session manager remarked that ESTCP had a large number of sites available for testing. This type of problem-oriented networking, and the associated crossing of sectoral boundaries and different phases of the innovation process, appears to be the norm at the SERDP Symposium. All of the interviewees echoed the view of the SERDP Symposium as providing crucial opportunities for networking and collaboration.

In contrast, the SERDP/ESTCP workshops focus on individual topics of interest. These meetings bring academics, industry, and military experts together to "examine the current state of the science and engineering on specific environmental problems."[30] These workshops help to determine what technologies and innovation paths seem most likely to succeed in advancing the DOD's environmental mission. The results of these workshops are often Workshop Reports, which summarize the priorities for research in a given field. For example, in October 2012, a SERDP workshop was convened by the program manager for SERDP's Environmental Restoration Program Area to discuss the long-term management of sediment contamination. Attendees examined the current state of technology for sediment contamination, discussed areas of interest for future studies, and then prioritized these technology needs for future DOD exploration.[31] Workshop findings help to keep SERDP and ESTCP on track with the needs of the DOD as a whole. Workshop findings also provide additional input into the annual SONs. While each technical area of interest for SERDP and ESTCP does not warrant an annual workshop, the major areas of research interest are revisited somewhat regularly in order to maintain a current view on a given topic's technical status as well as the institutional needs surrounding that technology or research area.

Box 3. A Comparable Program in the Department of Energy

The Office of Science and Technology (OST) within DOE's Environmental Management Program (DOE-EM) was dedicated to innovative remediation technology research, development and deployment (RD&D). OST existed from 1989 to 2003, and was funded at roughly $300 million per year.[32] Much like SERDP and ESTCP, OST's mission was to enhance innovation for environmental remediation, which it pursued by funding research and development topics ranging from basic science all the way through large-scale testing. During its first 10 years, OST funded over 700 projects. A detailed assessment by the Government Accountability Office (GAO, 1998) concluded that while much scientific data was produced by OST, the office was unable to function as a successful RD&D department due to several failings in its innovation methodology. A 1999 report by the National Research Council's (NRC's) Board on

Radioactive Waste Management (BRWM) confirmed the findings of the GAO report.[33] The GAO and NRC reports highlight several points of failure in the OST model.[34]

- Poor mission focus:
  - During its 14-year lifespan, OST was continually being reinvented, reorganized, and redirected both by its own management and by DOE-EM.
- Lack of engagement with end users:
  - GAO and numerous other agencies highlighted lack of user engagement as a major issue. By failing to provide solutions that were site- or organization- specific, OST managed to disenfranchise its end users and perform poorly in transferring its findings to real-world settings.
- Poor accountability for research funds:
  - OST failed to keep track of research funding, project results, or technical deployments in any meaningful way, thus making it extremely difficult to avoid redundant research or flag and either terminate or improve failing projects.
- Too strong a focus on bench science:
  - Only 1 in 5 projects funded by OST resulted in a meaningful demonstration or real world deployment. Of those that were tested, very few were ever deployed more than once.

Notably, each of the areas of weakness or failure identified at OST by GAO were highlighted as particular areas of strength and success at SERDP and ESTCP by all of those interviewed for this chapter, by the publicly available literature on SERDP and ESTCP, and by my own observations. Of particular note is the focus on bench science and the deployment rate (how often a given research project reached field testing) discussed in the 1998 GAO study. The same study compares OST's deployment rate to that of ESTCP, which was at that point only two years old. Even in its infancy, ESTCP managed to achieve field deployment for about 38% of its completed projects, which was comparable to OST's when projects at similar levels of development were compared. I reviewed all currently completed groundwater remediation projects undertaken by ESTCP since its inception to calculate their current deployment rate. Counting projects that have been field tested, had pilot plant runs, or have experienced full deployment, ESTCP has a current deployment rate of 79% in the area of groundwater remediation. Unfortunately, there is no current method in place for determining how often those technologies were deployed past their initial field/pilot tests. However, the high deployment rate for ESTCP is strongly indicative of successful innovation strategies, and contrasts markedly with OST's record.

Conclusion and Discussion

In seeking to fulfill its mission of advancing DOD's capability to better address its many environmental problems, SERDP and ESTCP have not only contributed significantly to environmental remediation on DOD and former DOD facilities through its research and field trials, but they have also catalyzed improved practice across the broad arena of environmental remediation in government and the private sector. ESTCP's 79% deployment rate of 79% in the groundwater remediation's area compares very favorably to that of NIST's ATP program, which achieved a 64% commercial prototyping rate and is widely considered a successful technology transfer program. SERDP has published guidance documents and advanced the state of the art of science for all groundwater remediation methods identified in a 2005 GAO study on the DOD's effectiveness in remediation technology development and deployment. Evidence for civilian spillover is circumstantial but strong: over 21 companies presented posters for SERDP's 2011 Symposium in the area of groundwater remediation. Each of these companies is involved both in civilian groundwater remediation, as well as working with government cleanup issues. While none of these companies explicitly point out on their websites that they use SERDP-developed technologies, they all use technologies SERDP and ESTCP have invested in for their military and civilian contracts.

Several notable qualities of the SERDP and ESTCP innovation methodology seem to be at the heart of this success. First and foremost, connecting stakeholders throughout the technological innovation process is essential when dealing with complex institutions and environmental problems. SERDP and ESTCP provide a managerial framework for their researchers, access to test sites, and a venue for publishing and sharing their data via the SERDP website and yearly symposiums. By engaging end users, academic researchers, industry partners, and regulators in their innovation process, SERDP and ESTCP have been able to produce an enormous amount of usable data, from basic science all the way through deployment costs and time-tables usable on real world remediation sites, as well as distribute that data to the organizations that can benefit from it. SERDP and ESTCP have a clear, transparent, and streamlined application process, clear guidelines for choosing which projects to fund, and a great deal of expertise on hand for supporting environmental remediation projects. This clear structure of reporting and accountability provides a managerial framework which encourages researchers, particularly those in academia who may not be used to strict deadlines and strong feedback on performance, to operate in a timely manner with a clear document trail for accountability purposes.

Part of the success of SERDP and ESTCP is certainly institutional. As part of the DOD, SERDP and ESTCP have access to numerous long term industry partners, remediation sites for testing, and a clear mission focus on environmental topics as a facet of national security. Furthermore, the DOD, besides having by far the largest budget of any U.S. government agency, also benefits from a certain amount of political insulation that surrounds security-related issues. Yet SERDP and ESTCP are not merely part of the DOD. At their core they may be funded by the DOD and focused on DOD-related problems, but they are in practice cross-institution programs with ties to the DOE, EPA, and numerous other organizations. Unlike similar environmental research programs, such as the DOE's former Office of Science and Technology, SERDP and ESTCP have made a point of involving all relevant stakeholders, from academic partners focused on basic research to large multinational firms engaged in remediation projects worldwide, in information exchange (perhaps better characterized as organizational learning) throughout the innovation process. Contractors and site managers are able to present their individual needs and seek out solutions at the yearly SERDP Symposium as well as receive information and training at workshops. Research results are plainly available online, easily searchable, and freely accessible to interested parties. Each of these facets of SERDP and ESTCP's innovation pathway points towards the importance of not merely having institutional ties and a mission focus, but the active maintenance of an innovation network as being the key to success in RD&D endeavors.

In addition to the institutional ties, internal management structures also play a large role in successful

RD&D activities. SERDP and ESTCP have built up clear structures for how research needs are determined, how individual proposals are evaluated and in the end chosen for funding, and how project progress is tracked. Part of the successful management of research projects also is shown by SERDP and ESTCP's breadth of research scale. They foster and understand the benefits of—and linkages among—a wide range of research, from small proof of concept SEED projects all the way to multi-site pilot plants.

Despite their success, SERDP and ESTCP do suffer from a common problem among government R&D groups, in that there is no easily accessible or derivable metric for success. As shown in the example of OST, ESTCP is clearly doing well in the category of deployment rate, but this one metric only tells part of the story. In order to develop a clear idea of SERDP and ESTCP's real value to DOD and the nation, new measures and information will be necessary. Doubtless the technologies developed by SERDP and ESTCP are being used in real world scenarios, as can be seen through the perchlorate case study. After all, military bases undertaking environmental remediation are currently hosting large scale operations designed by ESTCP. However, the lack of easily accessible data on how much remediation has been done, where it has been done, and what technologies were used at what price, make it very difficult to determine in the role that SERDP and ESTCP actually have played. Further studies on innovation pathways should be pursued to discover how to accurately track innovation success and failure. If the metric of deployment rates is seen to be useful for organizations like ESTCP, which likely seems the case, there needs to be a clear method of data collection for determining deployment rates. While the need for information on what technologies have been developed by SERDP and ESTCP and what companies are deploying them on which sites would seem to be obvious, none of my interviewees were able to point to a source for such details. Furthermore, my own fairly exhaustive search of publicly available information also failed to discover any readily available sources for determining deployment rates for R&D institutions. In total, there are many strong reasons to conclude that SERDP and ESTCP are remarkably effective programs for advancing environmental remediation innovation, and that the lessons of this success should be absorbed by other organizations seeking to enhance innovation to address complex problems in the environmental and other spheres. Nonetheless, additional, and more direct, information and measures that link SERDP and ESTCP's exemplary efforts to actual environmental remediation outcomes would make this case more compelling still.

---

[1] 10 U.S. Code § 2901 - Strategic Environmental Research and Development Program.

[2] Defense Environmental Research Program, "Defense Environmental Restoration Program Annual Report to Congress-Fiscal Year 1995," available at: http://www.denix.osd.mil/arc/DERPFY1995.cfm

[3] U.S. Government Accountability Office, *Nuclear Waste: Further Actions Needed to Increase the Use of Innovative Cleanup Technologies* (Washington, DC: U.S. GAO/RCED-98-249, Sept. 1998).

[4] Ibid.

[5] Jeanne Powell and Karen Lellock, *Development, Commercialization, and Diffusion of Enabling Technologies: Progress Report* (Washington, DC: U.S. Department of Commerce, Apr. 2000).

[6] U.S. Government Accountability Office, *Groundwater Contamination: DOD Uses and Develops a Range of Remediation Tech-*

nologies to Clean Up Military Sites (Washington, DC: U.S. GAO-05-666, June 2005).

7 Ibid.

8 U.S. Government Accountability Office, *DOD's New Environmental Security Strategy Faces Barriers* (Washington, DC: U.S. GAO/NSIAD-94-142, Sept. 1994).

9 Rudy Abramson, "Nunn Sees Military Role in Aiding Environment," *Los Angeles Times* (29 June 1990).

10 "Council and Executive Working Group," SERDP website, available at: https://serdp.org/About-SERDP-and-ESTCP/About-SERDP/Council-EWG

11 "Scientific Advisory Board," SERDP website, available at: https://serdp.org/About-SERDP-and-ESTCP/About-SERDP/Scientific-Advisory-Board

12 "SERDP Solicitations," SERDP website, available at: http://www.serdp-estcp.org/Funding-Opportunities/SERDP-Solicitations

13 Andrea Leeson, "Investment Strategy in SERDP and ESTCP," Federal Remediation Technologies Roundtable workshop presentation (May 2006).

14 "About SERDP," SERDP website, available at: https://serdp.org/About-SERDP-and-ESTCP/About-SERDP

15 "About ESTCP," SERDP website, available at: https://serdp.org/About-SERDP-and-ESTCP/About-ESTCP

16 "DNAPL Source Zones," SERDP website, available at: http://www.serdp-estcp.org/Featured-Initiatives/Cleanup-Initiatives/DNAPL-Source-Zones

17 "Managing the Environmental Impact of Perchlorate," SERDP website, available at: http://www.serdp-estcp.org/Featured-Initiatives/Cleanup-Initiatives/Perchlorate

[18] "Perchlorate," U.S. Environmental Protection Agency website, available at: http://water.epa.gov/drink/contaminants/unregulated/perchlorate.cfm#tabs-2

[19] California Assembly Bill 826, "The Perchlorate Contamination Prevention Act," 2003-2004 Regular Session (Feb. 2003).

[20] U.S. Government Accountability Office, *Perchlorate: Occurrence Is Widespread but at Varying Levels; Federal Agencies Have Taken Some Actions to Respond to and Lessen Releases,* (Washington, DC: U.S. GAO-10-769, Aug. 2010).

[21] Ibid.

[22] "SERDP Solicitations," SERDP website, available at: http://www.serdp-estcp.org/Funding-Opportunities/SERDP-Solicitations

[23] "Generation of Biodegradation-Sorption Barriers for Munition Constituents, ER-201123," SERDP website, available at: http://www.serdp-estcp.org/Program-Areas/Environmental-Restoration/Contaminated-Groundwater/Emerging-Issues/ER-201123

[24] Andrea Leeson and Hans Stroo, "SERDP & ESTCP: Environmental Restoration Research Program," workshop presentation (July 2008).

[25] U.S. Government Accountability Office, *Groundwater Contamination: DOD Uses and Develops a Range of Remediation Technologies to Clean Up Military Sites* (Washington, DC: U.S. GAO-05-666, June 2005).

[26] U.S. Government Accountability Office, *Nuclear Waste: Further Actions Needed to Increase the Use of Innovative Cleanup Technologies* (Washington, DC: U.S. GAO/RCED-98-249, Sept. 1998); *Incinerating Cash: The Department of Energy's Failure to Develop and Use Innovative Technologies to Clean up the Nuclear Waste Legacy* (Washington, DC: U.S. House of Representatives Staff Report, Oct. 2000).

[27] William Major, Mark Kram, Louise Parker, Tim McHale, and Joel Michelson, *Demonstration/Validation of Long-Term Monitoring Using Wells Installed by Direct-Push Technologies* (Port Hueneme, CA: Naval Facilities Engineering Command, Technical Memorandum 2410-ENV, June 2009).

[28] Federal Remediation Technologies Roundtable, *Guide to Documenting and Managing Cost and Performance Information for Remediation Projects*, Revised Version (Washington, DC: U.S. EPA 542-B-98-007, Oct. 1998).

[29] "Technology Screening Matrix," Federal Remediation Technologies Roundtable website, available at: http://www.frtr.gov/scrntools.htm

[30] "Conferences & Workshops," SERDP website, available at: http://www.serdp-estcp.org/News-and-Events/Conferences-Workshops

[31] Tim Thompson, Steve Ells, Marc Greenberg, Andrea Leeson, Victor Magar, Charles Menzie, Cara Patton, Michael Pound, Danny Reible, Deanne Rider, Jason Speicher, and John Wakeman, *SERDP and ESTCP Workshop on Research and Development Needs for Long-Term Management of Contaminated Sediments* (Washington, DC: SERDP/ESTCP Workshop Report, Oct. 2012).

[32] National Research Council, *Technologies for Environmental Management the Department of Energy's Office of Science and Technology* (Washington, DC: National Academies Press, 1999).

[33] Ibid.

[34] U.S. General Accounting Office, *Nuclear Waste: Further Actions Needed to Increase the Use of Innovative Cleanup Technologies* (Washington, DC: U.S. GAO/RCED-98-249, Sept. 1998).

3

AERODERIVATIVE GAS TURBINES*

Travis Doom

Introduction

For more than half a century, gas turbine engines pioneered for military jet fighters have hung under the wings of commercial airliners. For nearly as long, manufacturers have built industrial gas turbines to drive electricity generators and pump oil and gas. Many gas turbines, large and small, are designed specifically for these industrial applications. Aeroderivative gas turbines on the other hand are used for these industrial applications but are adapted directly from existing aircraft engines. Aeroderivative gas turbines emerged in the late 1960s with unique performance attributes in comparison to the existing industrial gas turbines. Aeroderivative units could startup more quickly for peak and emergency electricity generation, they weighed less and had a smaller footprint that was ideal for offshore platforms, and their higher efficiency and simplified installation and maintenance saved money for pipeline operators.

* This case study was supported and previously published by the American Energy Innovation Council at: http://americanenergyinnovation.org.

In the mid-1980s independent power producers began using aeroderivative gas turbines for combined heat and power generation, also called cogeneration. In this configuration, the exhaust heat of the gas turbine is used to produce steam to directly heat a building or industrial process. Aeroderivative gas turbines can convert 40 percent of fuel energy into electricity; when configured for cogeneration, system efficiency can exceed 80 percent, as far less of the fuel's chemical energy is lost as unused heat. Aeroderivative gas turbines are also being used to balance the integration of variable power sources, like wind and solar, into the electricity grid.

In 2010 the industrial gas turbine market—including aeroderivative gas turbines—was estimated to be $15.6 billion worldwide.[1] The electricity generation sector accounts for $12.9 billion of the total production value and is undergirded by the $2.2 billion mechanical drive sector, which includes applications for oil and gas production. The marine sector accounts for the remaining $0.5 billion in production value, where aeroderivative models are used to power the world's navies, fast ferries, and luxury cruise ships.

In the United States, over 20 percent of electricity generated in 2011 was powered by gas turbines, nearly all fueled by natural gas.[2] Gas turbines are increasingly important for U.S. electricity generation, with the Energy Information Administration projecting that natural gas generation will increase to 27 percent of all generation in 2025 and 30 percent in 2040."[3]

Most electricity generated by natural gas come from large industrial gas turbines designed specifically for that purpose, not aeroderivative units. These larger, heavy-duty or frame-type gas turbines for baseload power can exceed 200 MW (megawatts) each, and large plants may configure several of these units together in combined cycle. In this configuration exhaust heat from one or more gas turbines drives a separate steam turbine to generate additional electricity, achieving an overall efficiency of around 60 percent.

Aeroderivative gas turbines have historically been limited to midsize units of roughly 18 to 65 MW each, reflecting the size of their parent aircraft engines. Yet while the capacity of aeroderivative units is small, these units add flexibility critical for sustaining large generating plants throughout the electricity grid. Moreover, aeroderivative units are well suited for highly efficient cogeneration, more flexible combined cycle plants, and in mechanical drive applications they remain essential to production and distribution of the oil and gas that fuels our energy needs.

Worldwide, aeroderivative units account for approximately 10 to 20 percent of new gas turbine capacity in recent years, totaling roughly 6 to 8 GW (gigawatts) per year. However, for mid-size units with capacities of 18 to 65 MW, aeroderivative units account for approximately two thirds of gas turbine capacity sold in recent years.[4] General Electric (GE), the leading provider of aeroderivative gas turbines, estimates having over 2,300 aeroderivative units in service for electricity generation worldwide, totaling 80 GW of capacity.[5] Rolls Royce projects that aeroderivative units used for electricity generation and oil and gas operations over the next 20 years will be worth $70 billion in sales plus $50 billion for associated services.[6]

The entire industrial gas turbine market is just over half the size of the commercial and defense aircraft gas turbine engine market, which in 2010 amounted to more than $26 billion worldwide.[7] The performance of aeroderivative gas turbines is largely enabled by the engineering successes sustained in the lucrative aircraft engine market. The role and impact of industry-government partnerships in advancing aircraft engine technology is widely appreciated and has directly impacted the evolution of aeroderivative gas turbines. Much of the core aircraft engine research and development (R&D) has doubled effectively as aeroderivative gas turbine R&D, as the most difficult and costly to develop components in aircraft engines are used in nearly identical form in aeroderivative gas turbines. For example, over the last decade, spending on aircraft engine R&D at GE has exceeded $1 billion per year.[8] By comparison, over the last four decades, spending on aeroderivative gas turbine R&D at GE is estimated somewhere above $2 billion in total.[9] GE, Rolls Royce, and Pratt & Whitney are the three manufacturers of aeroderivative gas turbines, and each of their models is descendent from a specific aircraft engine in their lineup.[10] As a result, tremendous gains in aircraft engine performance, led by the military and bolstered by the rise of commercial aviation, have resulted directly in improvements for the aeroderivative gas turbines used throughout our energy system.

How Gas Turbines Work

Aeroderivative gas turbines, like all gas turbines, use a continuous intake of air and a continuous injection of fuel to create a hot, pressurized gas flow that expands through the turbine. In the compressor, a series of rotating blades pressurizes the incoming air in stages; this pressurization heats the air. In the combustor, chemical energy from the burning fuel adds far more heat. The hot, pressurized gas expands through the turbine blades and rotates the shaft that drives the compressor at the front of the engine, continuing the cycle. From this basic configuration, the remaining energy not used to drive the compressor can be captured in useful ways for various applications. In aircraft engines, the hot exhaust gas passes through additional turbines to rotate a shaft driving a propeller or fan that provides most of the aircraft's thrust. Aeroderivative and other industrial gas turbines work in a similar way, adding more turbines to extract energy from the hot exhaust gas to power a shaft. For industrial applications, the shaft is connected to an external electricity generator, a pump, or a ship's drivetrain.

History and Development

Gas turbine development arose in the early 1940s from the military demand for high-performance aircraft. This demand existed because of performance limitations inherent to the piston engine-propeller aircraft of the era, and because nations embroiled in conflict faced this common challenge. In the U.S., gas turbine innovation for aircraft superiority gained urgent and strategic focus under its national security mission, and the U.S. military fostered the competition necessary to explore a variety of initial designs. As a result, innovation accelerated through the 1940s and 1950s.

Gas turbine development in the U.S. began with several concurrent efforts. GE built a design borrowed from Frank Whittle (the English inventor of the turbojet engine in the 1930s) for the Air Force, and Westinghouse built their own design for the Navy.[11] The Navy also helped to pull Pratt & Whitney into the competition by contracting them to manufacture Rolls Royce gas turbines. Pratt & Whitney quickly maneuvered to build and sell their own designs to the Air Force as well.[12] Early efforts were underway by the National Advisory Committee for Aeronautics (the predecessor to NASA) to study compressor aerodynamics, but it was the unsurpassed multitude of engine orders in the 1950s that motivated competing firms to develop advanced designs.

Amid this competition, Pratt & Whitney decided to pursue a high-compression gas turbine that would significantly improve fuel efficiency and power for the B-52 bomber. State-of-the-art compressors at the time reached a pressure ratio around 6 to 1, and Pratt & Whitney was aiming to more than double that. The major design change devised by Pratt & Whitney to achieve this radically higher pressure ratio let the compressor operate in two sections on independent shafts at different speeds. This twin-shaft configuration, enabling stable operation from startup to full throttle, led to the highly successful J-57 engine.[13] Pratt & Whitney built more than 20,000 J-57 engines for the Air Force and Navy between 1951 and 1965 and less than 1,000 J-57 engines for the nascent commercial aviation jet fleet.[14] The J-57 soon evolved into the more powerful J-75 engine.

GE devised an entirely different solution to ensure a stable, high-pressure compressor for their J-79 engine to power the B-58 bomber. The GE variable stator design used a single shaft and mechanized parts of the compressor to adjust the geometry as airflow conditions changed.[15] Vigorous demand led GE to create fifteen new engine variations during the 1950s and build more than 17,000 J-79 engines for the military.[16] The Pratt & Whitney J-75 and GE J-79 would become the parent engines for their first aeroderivative gas turbines. High-pressure compressors were the heart of these new engines, lending aeroderivative gas turbines their exceptional simple-cycle efficiency.

The drivers of gas turbine technology development shifted in the 1960s. The military began growing a technology base that, separate from enormous procurement programs, would sustain innovation in the decades ahead. The efforts of the 1940s and 1950s had resulted in innovative engine designs like the J-57 and J-79, but had also incurred huge costs with numerous, inexplicable problems, as described in one study of energy innovation at the Department of Defense:

> *Machinery when built did not behave as predicated. Nor could analytical methods provide much guidance on what to change. The difficulties encountered with a particular engine design might be overcome, but all too often the engineering*

group, in the absence of insight based on physical understanding, had no choice but to proceed by trial and error. Research was the necessary route to findings that could be generalized.[17]

A concept took hold within the Air Force Propulsion Lab to conduct gas turbine research, not by constructing and testing complete engines, but instead by improving the core components. This "building block" approach was intended to develop the compressor, combustor, and turbine as a unit suited to a range of future engines. An Air Force report explains why this was an organizational, as well as technical, innovation:

[T]he lab had to convince the various contractors that this was the way forward, because the idea was radical at the time and its success was not assured or embraced. For contractors, this was a conceptual leap ... several years and millions of dollars in contracts would not result in a particular contractor ending up with a singular, superior engine.[18]

The reasons this concept took shape after 1959 were simple: the Aeronautical Systems Division, not the Air Force Propulsion Lab, was given responsibility for new engine programs in the U.S. Air Force. With much Air Force funding also being consumed in the Space Race, the Turbine Engine Division within the Air Force Propulsion Lab needed a more focused approach to assert their relevance.[19] This organizational shift was not a one-off effort, but rather part of much broader institutional change. The Air Force was learning to manage the development of an astounding array of complicated technologies. Gas turbine engines, like ballistic missiles and command and control computer systems, required new approaches to engineering and management.[20]

While the Air Force may have provided the initial nudge for the "building block" approach, within industry the shift towards a stronger technology base was further incentivized by the rise of commercial aviation. GE used the opportunity to take the same core components validated for military transport aircraft to launch their commercial engine line. In 1979, GE described this alignment as the "most significant business/technology achievement to date in GE aircraft engine history."[21] Big profits were on the horizon for firms that understood gas turbine technology well enough to balance tradeoffs and improve performance along all dimensions: power, efficiency, reliability, maintainability, and cost.

Aeroderivative gas turbines benefitted for decades from an uninterrupted stream of incremental innovations flowing from their parent aircraft engines. For example, GE continually improved its LM2500 aeroderivative model and, in 1990, introduced the significantly more powerful and efficient LM6000, which was based on an aircraft engine that "embodied all the aerodynamic and materials refinements since its 1970-predecessor."[22] These refinements came from both the engineering science supported by the Air Force and the millions of flight hours accumulated in the thriving commercial aviation sector.

The U.S. Navy transitioned its fleet of ships from steam to aeroderivative gas turbine power beginning in 1970, following experiments in the 1940s and a concerted development effort throughout the 1950s and 1960s.[23] Not unlike the ongoing efforts with aircraft engines at the time, much of the learning first took place by repairing and improving marine gas turbines already in service.[24] Because initial testing proved inadequate at mitigating problems in development engines throughout the fleet, the U.S. Navy implemented more rigorous rating and qualification standards for marine gas turbines. Understanding, for example, how the severity of corrosion and oxidation varied for different alloys and coatings under realistic operating conditions required extensive empirical efforts: "The life of ultra-high temperature materials in gas turbine engines (shipboard or aircraft) is dependent on a complex combination of temperature-stress-environment-time variable fields."[25]

In 1968 Pratt & Whitney and GE aeroderivative units underwent extensive, side-by-side testing for the U.S. Navy.[26] Although some operating conditions, like high salinity, are unique to marine applications, the materials engineering knowledge cultivated through these efforts was valuable more generally for adapting aircraft engines for industrial service in harsh environments. For example, the high sulfur content in lower quality, non-aviation fuels presented service-life concerns for aeroderivative gas turbines installed in ships—a problem common to gas turbines on pipelines and at peak generator stations. The U.S. Navy played an important role early on to help advance aeroderivative gas turbines for more diverse uses, both by demanding a robust product and supporting the engineering and testing to further demonstrate durability.

Since the late 1940s, firms were leveraging their knowledge and investment in aircraft engines to explore industrial gas turbine applications. Westinghouse, for example, built a pair of gas turbines that powered a Baldwin locomotive in 1948. When the experimental locomotive was scrapped, Westinghouse resold the gas turbines to power a peak-electricity generating plant in Kansas and an industrial air compressor in Virginia.[27] Both GE and Westinghouse stood on many years of experience building steam turbines, but it was the very different challenge of designing military aircraft engines that provided a basis for their industrial gas turbine businesses. Even Siemens, who exclusively builds industrial gas turbines and who also relied in part on steam turbine experience, traces their history to a German aircraft engine built in the 1940s.[28]

Industrial gas turbine businesses utilized the aerodynamics developed for aircraft engines, but the industrial versions evolved apart from the design constraints unique to aviation. Industrial gas turbines were not engineered to minimize size and weight. Durability and affordability was prioritized for industrial gas turbine customers, which resulted in components with greater wall thickness and different material choices. Industrial gas turbines also tended to use moderate pressure ratios and rely on auxiliary devices like heat exchangers to improve efficiency. The niche for aeroderivative gas turbines that emerged in the 1960s was shaped by these design differences and strengthened by rapidly advancing aircraft engine performance.

When manufacturers started repurposing aircraft engines for industrial applications, aeroderivative gas turbines had 15 to 25 MW capacities with efficiencies near 30 percent. Pipeline operators could more easily transport, install, and maintain the lightweight and compact aeroderivative gas turbines in remote locations, and offshore platforms often needed the higher power in a smaller footprint that aeroderivative gas turbines offered. Blackouts in the U.S. and Europe in the 1960s also spurred the use of aeroderivative gas turbines for emergency and peak electricity generating applications. Aeroderivative gas turbines, with their thinner, lighter-weight components, could ramp up to operating temperature more quickly than other industrial gas turbines, a valuable feature for emergency and peak power applications.[29] Cogeneration, another major application using aeroderivative gas turbines, took off in the mid-1980s following a series of policies deregulating energy markets.[30] The Public Utility Regulatory Policies Act of 1978 first prompted the expansion of cogeneration capacity, although it was at first limited mostly to coal-fired steam turbines.[31] After the price of natural gas resettled following deregulation completed in 1985, and after the Power Plant and Industrial Fuel Use Act was repealed in 1987 following newly available reserves, independent power producers became major customers for the aeroderivative gas turbine businesses.[32]

The momentous shift in natural gas price and availability, along with environmental concerns and changes in electricity markets, also resurrected the demand among utilities in the U.S. for larger, baseload industrial gas turbines.[33] Industrial gas turbine manufacturers were eager to infuse the latest aircraft engine technology in their products, but only GE could source this knowledge internally from their own aviation division (as Westinghouse had exited the aircraft engine business in the 1950s). In the 1990s, Siemens and Westinghouse managed alliances with aeroderivative turbine developers Pratt & Whitney and Rolls Royce, respectively. Although these industrial gas turbines would retain their own heavy-duty designs, the engineering tools needed for example to improve compressor blade efficiency and enhance turbine cooling were far better developed within the firms building aircraft engines.[34]

One exemplary program, the GUIde (Government Agencies, Universities, and Industry) Consortium, highlights the contours of the aircraft engine technology base and details the kinds of partnerships that have long advanced innovation for gas turbines. First convened in 1991, the GUIde Consortium set as its goal mitigating certain low-amplitude, high-frequency vibrations that cause compressor blades and other components to fail. This type of failure, known as high cycle fatigue (HCF), became increasingly problematic:

> *As the decade of the 90s opened HCF became a more notable problem as the B-1 engine experienced two spectacular failures, one of which led to the engine separating from the aircraft ... By the mid-1990s HCF issues became the dominant failure mode for fighter engines in the USAF. With tensions high on the Korean peninsula, specially equipped F-16s with the mission to suppress enemy air defenses were grounded due to an HCF issue ... Also during this time HCF began to appear in commercial engines to a lesser extent than in mili-*

tary engines but with severe consequences to the manufacturer's development programs and revenue service for airlines.[35]

The GUIde Consortium was built around the reality that HCF was an industry-wide problem. HCF had persisted because the difficulty of the problem was beyond the reach of individual firms who had limited funds to improve their own design systems for predicting vibration.[36] The words of a leading expert from the Air Force capture the difficult nature of the problem that necessitated a cooperative approach among otherwise competitive businesses:

I can recall many meetings where, as a group of technical experts, we went through a systematic analysis of the conditions leading to an HCF failure and can prove, through existing data, knowledge, and analysis that a failure could not have occurred. Only the failed parts in our hands were able to convince us of our inability to completely describe the event accurately.[37]

The GUIde Consortium was structured with center directors at Carnegie Mellon University and Purdue University, with a steering committee of voting members, one each from the Air Force, NASA, and six participating companies. Industry provided approximately half of the funding for the six research projects approved by the steering committee, which were conducted mostly at universities, each at a level around $100,000 per year for four years. This arrangement allowed the individual firms to leverage their investment approximately 10 to 1, and it allowed the multidisciplinary problem to be effectively divided into key research areas, such as structural dampening and unsteady aerodynamics.[38] Although HCF did not afflict the existing generation of aeroderivative gas turbines, other industrial gas turbine manufacturers, including Siemens and Mitsubishi, joined the GUIde Consortium. These firms needed better engineering tools to ensure their newest machines would not experience damaging vibrations operating across a range of speeds with more aggressive compressors—the same tools that the Air Force and its contractors needed to restore reliability in their fleet of fighter aircraft engines. NASA also joined the GUIde Consortium, seeking better tools to predict vibration in rocket engines.

University researchers in the GUIde Consortium needed data to develop models of the phenomena causing HCF, and the participating companies needed models that could be practically implemented in their existing design systems. Turbine performance data, such as force and flow measurements and detailed component geometries, was transferred from companies to researchers through subcommittees overseeing individual projects. The subcommittees facilitated closer interaction between the technology developers and the technology users, providing opportunities to clarify the objectives and validate the new tools. The GUIde Consortium also provided the structure needed to share and analyze complete data sets from expensive rig tests at Air Force and NASA facilities. Additionally, review meetings and conferences strengthened the small community of dedicated experts.[39]

The GUIde Consortium gained momentum in its first few years, and the fifth iteration of the program began in October 2013. Between 1995 and 2005 the Air Force led a comprehensive, highly successful $400 million initiative to mitigate HCF, with the GUIde Consortium program continuing as one part. The field engine inspection workload for HCF declined by over 90 percent, and the proportion of engine mishaps resulting from HCF declined from 54 to 7 percent.[40] The GUIde Consortium example illustrates the kinds of goal-oriented, incremental innovations that improve aircraft engine performance while further strengthening the knowledge base for industrial gas turbines, including aeroderivative models.

Current Status of the Technology

Aeroderivative gas turbines today remain a leverage business, enabling GE, Pratt & Whitney, and Rolls Royce to reach other markets while minimizing the cost and risk associated with new technology development. These firms utilize both methods and components developed from their lucrative commercial and military aircraft engine businesses. Much of this knowledge has diffused through the industry-government partnerships and alliances discussed above, ultimately improving the gas turbine technology for electricity generation and other industrial applications.

While the focus of this case study has been aeroderivative gas turbines, all industrial gas turbines have benefited from the aircraft engine technology base. In the 1990s, the firms that most effectively sourced aircraft engine R&D—GE internally and Siemens externally from Pratt & Whitney—became market leaders. Of course, the success of a firm's gas turbine business is based on far more than R&D; GE and Siemens have also proven adept at responding to technical mishaps through effective marketing and sales support.[41]

Many gas turbines today blur the boundaries that generally distinguish aeroderivative gas turbines. Siemens, for example, has acquired the knowledge needed to optimize certain designs for lower weight and compact footprints with faster startup times and higher simple cycle efficiency. GE now builds a gas turbine that couples an aeroderivative core with their larger, heavy-duty turbine components in a unit that produces 100 MW, a power level much higher than could be obtained by simply repurposing an aircraft engine. Moreover, as electric grid operators increasingly value flexible generation assets to deal with new challenges, aeroderivative gas turbines are being incorporated into a variety of combined cycle gas turbine configurations, thereby providing fast-start and cycling capabilities while retaining the overall efficiency of an integrated system. With power generation from natural gas and variable renewable sources on the rise in the U.S., the use of industrial gas turbines will increase and the competition to offer customers a full range of solutions will intensify.

Lessons Learned

The history of aeroderivative turbine development underscores the power of mission-oriented and demand-driven R&D. Procurement not only established the business case for private R&D efforts, but also ensured direct feedback from customers to the carefully managed industry development efforts. That said, the lessons for aeroderivative gas turbines are rooted in the unique context and history of aircraft engine development. It is critical to remember that the conditions created by World War II in the 1940s, the Korean War in the 1950s, and the Cold War in the years beyond are far from a practical context for budget-constrained energy technology policy planning today. One historian remarks:

> *Viewed from a distance, the development of jet propulsion in the U.S. may appear to have been a chronicle of progress through skillful management of technology and organization. Examined closely, it stands rather as a shining example of non-linear, irrational, uncertain, multi-lateral, and profoundly passionate technological and business practice, yielding success not through planning but through dogged determination, a certain indifference to failure (which secrecy aided), and massive expenditures of public funds.*[42]

This excerpt perhaps best characterizes the earliest decades of gas turbine development. Lessons for innovation policy today may instead reside in the more recent, carefully planned and managed programs exemplified by the GUIde Consortium. However, here too consideration is needed for the context unique to military technology development. The government role in advancing gas turbine performance was motivated by national defense objectives, not commercial or environmental aims.

In the 1990s, the short-lived Technology Reinvestment Project (TRP) tried to overcome this divide, bridging military and commercial needs. Despite sharing key attributes with the successful GUIde Consortium, including a cost-sharing structure and industry led teams, TRP encountered serious difficulties that led to the program's demise. Tradeoffs between military and commercial requirements strained technology programs funded through TRP, and the partnerships threatened to undermine participating military contractors' competitive advantage in the defense acquisition process.[43] In contrast, features inherent to gas turbine engines have allowed program managers to navigate around such difficulties. Most military and commercial aircraft engines are not critically different, and gas turbines are sufficiently complicated that multidisciplinary partnerships do not hinder competition among engine builders. The core competency for these firms is the ability to integrate applied research into a useful design system, carried forward into a demonstrator engine, while minimizing risk and cost and following documentation and contractor guidelines through to manufacturing—an altogether enormous task.

The lesson for energy innovation is that aligning research with customer priorities and user needs is not a downstream or translational activity, but rather a central part of a successful R&D process. This remains a defining feature of the end-to-end military enterprise and is vital for pushing forward innovation in other domains. Gas turbine aircraft engines have been central to the culture of aviation surrounding the Air Force, from jet fighters to long-range bombers and huge transport aircraft. Not all energy technologies simultaneously serve a public good and provide clear commercial value. Nor do all energy technologies have a proven architecture suited to incremental gains through engineering science and operational learning. The challenge for energy innovation is to craft policy that strengthens the public goods framework for technologies that can be built, improved, and will ultimately contribute to national goals for clean, reliable, and affordable energy.

---

[1] Lee S. Langston, "Powering Ahead," *Mechanical Engineering* 133, no. 5 (2011): pp. 30-33.

[2] This percentage includes the electricity produced by steam turbines running in combined cycle with gas turbines. See: "Form EIA-923 detailed data," Energy Information Administration website, available at: http://www.eia.gov/electricity/data/eia923/

[3] U.S. Energy Information Administration, *Annual Energy Outlook 2013 with Projections to 2040* (Washington, DC: U.S. EIA, DOE/EIA0383, Apr. 2013).

[4] "BY THE NUMBERS: 25% more GT capacity will be ordered for electric generation in the US this year than in 2012," *Combined Cycle Journal* website, available at: http://www.ccj-online.com/by-the-numbers-25-more-gt-capacity-will-be-ordered-for-electric-generation-in-the-us-this-year-than-in-2012/

[5] "Grounded: How GE's Flightless Jet Engines Power the World," GE Reports website (2012), available at: http://www.gereports.com/grounded-how-ges-flightless-jet-engines-power-the-world/

[6] Rolls-Royce, *Rolls-Royce Holdings plc Annual Report 2011* (London, UK: Rolls-Royce, 2012), available at: http://www.rolls-royce.com/Images/RR_full_AR_2011_tcm92-34435.pdf

[7] Langston, "Powering Ahead."

[8] Gilbert H. Badeer, "GE's LM2500+G4 Aeroderivative Gas Turbine for Marine and Industrial Applications," GE Technical

Paper: GER-4250 [09/05], available at: http://site.ge-energy.com/prod_serv/products/tech_docs/en/downloads/ger4250.pdf

[9] Edward Wacek and Warren Ferguson, "Application of the latest GE aeroderivative gas turbine technology," paper presented at the 2009 Industrial Applications of Gas Turbines symposium, available at: http://www.iagtcommittee.com/downloads/2009papers/203%20paper%20final.pdf

[10] Pratt & Whitney recently divested their aeroderivative business to Mitsubishi Heavy Industries.

[11] A. J. Scalzo, R. L. Bannister, M. DeCorso, and G. S. Howard, "Evolution of Westinghouse Heavy-Duty Power Generation and Industrial Combustion Turbines," *Journal of Engineering for Gas Turbines and Power* 118, (Apr. 1996): pp. 316-330.

[12] Jack Connors, *The Engines of Pratt & Whitney: A Technical History* (Reston, VA: AIAA Press, 2010).

[13] Ibid.

[14] Philip Scranton, "Urgency, uncertainty, and innovation: Building jet engines in postwar America," *Management & Organizational History* 1, no. 2 (2006): p. 140.

[15] Bernard L. Koff, "Gas Turbine Technology Evolution: A Designer's Perspective," *Journal of Propulsion and Power* 20, no. 4 (2004).

[16] General Electric, *Seven Decades of Progress: A Heritage of Aircraft Turbine Technology* (Aero Publishers, 1979).

[17] John A. Alic, "Defense Department Energy Innovation: Three Cases," in *Energy Innovation at the Department of Defense: Assessing the Opportunities* (Washington, DC and Boston, MA: Consortium for Science, Policy & Outcomes and Clean Air Task Force, Mar. 2012): p. 16.

[18] B. Oder, J. Aldridge, J. Marchese, R. Duffner, and S. Christy, *Breakthrough Technologies Developed by the Air Force Research Laboratory and Its Predecessors* (Wright-Patterson AFB, OH: Air Force Research Laboratory History Program, 2005): p. 132.

[19] Ibid.

[20] Stephen B. Johnson, *The United States Air Force and the Culture of Innovation: 1945-1965* (Washington, DC: Air Force History Museums Program, 2002); and John A. Alic, *Trillions for Military Technology: How the Pentagon Innovates and Why it Costs So Much* (New York, NY: Palgrave Macmillan, 2007).

[21] General Electric, *Seven Decades of Progress.*

[22] Raymond F. Patt, "GE Industrial Aeroderivative Technologies," in *Proceedings of ASME Turbo Expo 1998: Power for Land, Sea and Air* (1998).

[23] David Shifler, Donald Hoffman, John Hartranft, Carla Grala, Louis Aprigliano, and Dan Groghan, "USN Marine Gas Turbine Development Initiatives Part I: Advanced High Temperature Materials," in *Proceedings of ASME Turbo Expo 2010: Power for Land, Sea and Air* (2010), p. 1071.

[24] William S. Hong and Paul D. Collopy, "Technology for Jet Engines: Case Study in Science and Technology Development," *Journal of Propulsion and Power* 21, no. 5 (2005): pp 769-777.

[25] Shifler, et al., "USN Marine Gas Turbine Development."

[26] Ibid.

[27] Scalzo, et al., "Evolution of Westinghouse."

[28] Ihor Diakunchak, Hans Juergen Kiesow, and Gerard McQuiggan, "The History of the Siemens Gas Turbine," in *Proceedings of ASME Turbo Expo 2008: Power for Land, Sea and Air* (2008).

[29] H. I. H. Saravanamuttoo, G. F. C. Rogers, H. Cohen, and P. V. Straznicky, *Gas Turbine Theory*, 6th ed. (Essex, UK: Pearson Education, 2009).

[30] Robert H. Williams and Eric D. Larson, "Aeroderivative Turbines for Stationary Power," *Annual Energy Review* 13 (1988): pp. 429-489.

[31] Richard F. Hirsch, *Power Loss: The Origins of Deregulation and Restructuring in the American Electric Utility System* (Cambridge, MA: MIT Press, 1999).

[32] Darian Unger and Howard Herzog, "Comparative Study on Energy R&D Performance: Gas Turbine Case Study," report prepared for Central Research Institute of Electric Power Industry (1998).

[33] Ibid.

[34] Anna Bergek, Fredrik Tell, Christian Berggen, and Jim Watson, "Technological capabilities and late shakeouts: industrial dynamics in the advanced gas turbine industry, 1987-2002," *Industrial and Corporate Change* 17, no. 2 (2008): pp. 335-392.

[35] Otha Davenport, "Final Report for USAF High Cycle Fatigue Program," Appendix B in Theodore Nicholas, *High Cycle Fatigue: A Mechanics of Materials Perspective* (Oxford, UK: Elsevier, 2006).

[36] Scott M. Richardson and Jerry H. Griffin, "The GUIde Consortium: Government, Universities, and Industry Working Together to Develop New Technologies," in *Proceedings of the 1995 International Gas Turbine and Aeroengine Congress and Exposition*, Houston, TX (1995).

[37] Theodore Nicholas, *High Cycle Fatigue: A Mechanics of Materials Perspective* (Oxford, UK: Elsevier, 2006).

[38] Richardson and Griffin, "The GUIde Consortium."

[39] Ibid.

[40] Davenport, "Final Report."

[41] Bergek, et al., "Technological capabilities."

[42] Scranton, "Urgency, uncertainty, and innovation."

[43] Linda R. Cohen, "Dual-Use and the Technology Reinvestment Project" in Lewis M. Branscomb and James I. Keller, eds., *Investing in Innovation: Creating a Research and Innovation Policy that Works* (Cambridge, MA: MIT Press, 1998); and Jay Stowsky, "The Dual-Use Dilemma," *Issues in Science and Technology* (Winter 1996).

4

THE EFFECTS OF THE DEPARTMENT OF DEFENSE ON THE PHOTOVOLTAIC INDUSTRY

Frank N. Laird

Introduction

The U.S. Defense Department (DOD) has been a major catalyst for technological innovation across numerous fields. Can it do the same for photovoltaics (PV)? The DOD could affect the PV industry through four routes, three economic and one symbolic. The first and most obvious route is simply through procurement of relatively mature PV technologies. DOD purchases have been important to other technologies as they provide a large customer and therefore a steady cash flow for the industry. Second, DOD could function as an early adopter for more cutting-edge PV technologies. Sometimes the military needs products with special performance characteristics for which it is willing to pay a premium, more than would a conventional civilian market. While early adopter purchases may not be large in an absolute sense, they can be important to infant industries or to new products in mature industries. Third, the DOD has a large research and development budget and could choose to use some of that to advance technological development for next-generation PV.

Fourth, DOD and military use of PVs could have an important symbolic effect. Such effects are more difficult to quantify, but are a significant part of moving these technologies into the social mainstream. Photovoltaics, and renewable energy in general, have a deeply politicized history and DOD use of these technologies is part of a trend of breaking away from that politicized past.

Economic Effects

The PV market is by now a highly segmented one, ranging from older technologies which the industry produces at the rate of gigawatts per year, to newer technologies that currently occupy small niche markets, to technologies that have not left the lab bench.[1] As Weiss and Bonvillian note, the most effective policies for promoting an energy innovation depend on the state of development of that technology along with the market conditions in which it is competing.[2] Therefore, the effects of DOD procurement or R&D spending will depend on the status of particular PV technologies.

At various times in the past, government agencies have been crucial as both funders of R&D and final customers for the resulting products for PV technologies at all the these stages, most notably the National Aeronautics and Space Administration spending on both R&D and procurement during the 1950s and 60s and the Department of Energy as a source of R&D funding beginning in the 1970s.[3] What role now does the DOD play?

Procuring Mature Technologies

The Defense Department is a customer for mature PV technologies, but not a large one compared to the size of the U.S. PV market as a whole, and even less so when considering the size of the global market. We can get a sense of the relative size of DOD markets for PV through two different measures, large installations and total sales of PV.

The Solar Energy Industries Association (SEIA) periodically updates a report on utility-scale solar installations, by which they mean ground-mounted solar plants larger than one megawatt (1 MW).[4] The list includes facilities that are completed, under construction, and in development, along with their capacity in MW.[5] Some of the projects that are under development may never be built, so the MW in that category is a bit speculative for both DOD and the total.

The results are striking. Projects undertaken by various components of DOD are only a small part of the total installed capacity, given below in MW. Procurement of large-scale PV facilities by DOD is on the order of 1% of existing facilities. While no doubt individual firms wish to get every sale they can, it is clear that DOD procurement does not drive this market as a whole.

DOD procurement could be more important than the simple numbers indicate if procurement policies were more stable. The financial incentives that the federal and state governments provide, especially the federal investment tax credit, along with some that come from utilities, are due to expire soon and there has been a long history of on-again, off-again federal incentives. If DOD procured PV without concern for those tax credits, then it could be a steadier customer for PV than utility projects that depend on them. However, the large DOD PV projects so far do in fact make use of these subsidies. In the case of large PV installations at Camp Pendleton (Marine Corps), Fort Carson (Army), and Nellis Air Force Base, the military did not buy the PV systems outright. Instead, all three facilities entered into contracts with private sector firms and local utilities that provided the financing for the PV systems. The military bases located the PV systems on their land and bought their output.[6] This means that the utilities and other investors could make full use of federal and other incentives. If this is the procedure that DOD uses for financing PV on its bases, that suggests that if and when the incentives expire, DOD procurement for large PV systems may decline as well. As large as the DOD budget is, any given base has constraints on the money it can spend on electricity.

These large facilities are not the only part of the PV market. When seen against the entire U.S. PV market, the DOD fraction looks even smaller. By the end of 2013, the United States had a total installed PV capacity (that is, systems that were operating) of 12.1 gigawatts (GW), dwarfing the DOD large-systems installed capacity of 60 MW.[7] Of this large total U.S. capacity, about two-thirds is not in utility-scale systems; most of the capacity is made up of systems smaller than 1 MW. SEIA estimates that there were 445,000 PV systems in the United States by the end of 2013. Even if DOD has smaller PV systems comprising about two-thirds of its total capacity, the same proportion as the rest of the country, its smaller systems would still give it only 180 MW out of 12.1 GW, or about 1.5% of the total.

We can further contextualize the DOD purchases by putting them into global perspective. In 2011 the United States was 7% of the global market for PV, which makes the DOD installations a fraction of a fraction.[8] The global size of the market is relevant both because U.S. manufacturers compete for international customers, and because PV developers who install systems in the United States use both imported and domestically-manufactured PV cells and modules when building systems. With its vanishingly small percentage of the global PV market and no utility-scale PV projects under development, it is clear that DOD procurement cannot drive this industry.

Table 1. Utility-Scale Solar PV in the U.S.

Status	Total MW	DOD MW	DOD %
Operating	4,995	61	1.2
Under Construction	3,356	1	0
Under Development	23,143	none	0
Total	27,295	62	0.2

Source: SEIA, Major Solar Projects in the United States (2014).

Procuring Advanced Technologies

The DOD can advance cutting-edge renewable energy technologies that are not yet mature and which have little or no presence in civilian markets, simply by purchasing them. The DOD demand for these technologies comes from the same motivation that has led it to procure other advanced technologies. Combat creates strenuous performance requirements for military equipment and the DOD is willing to pay a premium for technologies that can fulfill those requirements. As such, DOD could be an important customer for that segment of the industry.

Assessing just how important DOD is to this segment of the industry is difficult due to data limitations and lack of data comparability. First, we do not know exactly how large DOD procurement is of these products. The General Accountability Office did a survey of renewable energy initiatives in the federal government, including the DOD, but that survey only counted the number of initiatives in each agency, it did not provide financial data for any of them.[9] The report lists three initiatives in DOD that would lead to procurement of advanced PV, two in the Marine Corps and one in the Army. Each of these initiatives includes efforts to provide portable PV systems to soldiers in combat environments. The goal is to reduce the amount of weight soldiers have to carry to power their systems and the fuel that forward bases require to generate the electricity that they need. Therefore, DOD procurement for these programs would target very light-weight and efficient systems, technologies like organic PV instead of conventional polysilicon PV (the mature and largest part of the industry). This procurement could make DOD an important customer for the advanced segment of the industry.

Publicly available reports do not indicate the size of military acquisitions of these technologies. But data about the size of the overall U.S. market suggest that DOD procurement would not have to be very large to make a significant impact on the market if it were focused on acquiring the latest innovations. The standard Energy Information Administration report on PV presents shipments by type, but does not separate out advanced PV such as those on flexible plastic backings or other very light-weight technologies.[10] If such advanced PV are only a small fraction of the market, then even modest DOD procurement could provide a significant boost, but available data do not allow us to assess whether this is the case.

Funding RDT&E

The DOD has a large budget for research, development, and demonstration, a set of categories that also often include testing and evaluation, or RDT&E. However, the amount that it spends on such work for PV is difficult to assess. The Consortium for Science, Policy & Outcomes project on energy innovation has combed through recent DOD budget documents and estimates that $142 million in recent years has been spent on RDT&E for PV (http://archive.cspo.org/projects/eisbu/atlas).

These funds are cumulative over FY 2009 through FY 2013 and include special one-time stimulus funds. At an average of $36 million per year, these funds comprise only a small fraction of the available federal funding for PV. By comparison, the major source of funding is the Energy Efficiency and Renewable Energy program in the Department of Energy. In FY 2013 that program received $150.6 million for PV R&D, plus $45.8 million and $29.6 million for R&D on PV systems integration and balance of systems, respectively.[11] This total of $226 million for a single fiscal year is far greater than the funds provided by DOD. While any research community is glad to receive a greater budget for its work, the fraction funded by DOD is unlikely to drive the efforts of the community.

However, like the case of procurement of advanced technologies, one cannot rule out a significant role for DOD funding of R&D. Even though DOD funds are not large in the aggregate sense, it may be the case that DOD applies them to crucial areas that other agencies neglect or applies them through more effective funding arrangements, allowing firms to develop these technologies more quickly than they otherwise could. The best way to assess such an effect would be to conduct several case studies of firms that specialize in these advanced technologies and inquire about the sources of the innovations that they use.

Solar Soldiers and Solar Symbolism

Difficult to measure, but potentially no less important, than the modes of investment discussed so far is the symbolic effect that DOD can have upon the development and deployment of PV. Bell Labs developed the earliest silicon PV and NASA provided an early and very high profile market for them by using them to power early satellites.[12] They were the epitome of high tech, used in the most visible and glamorous of government programs. However, the absolute amount of energy they generated was small and they were far too expensive to use for generating bulk power that could compete with fossil-fuel power plants.

It was not until the energy crisis in the 1970s, as policy makers and the public confronted severe shortages and consequently rapidly rising prices for fossil fuels, that government policy and parts of the professional literature began to consider PV as a possible alternative source of energy. The most visible feature of the energy crisis was the oil shortage and stunning increase in the cost of gasoline, triggered by OPEC's price increases and, for a few months in 1973, embargo of the United States and several European countries. In addition, the prices of coal had risen and there had been some coal shortages, mainly caused by labor disputes, prior to the oil embargo.[13] The federal government dramatically increased its funding for renewable energy sources, including PV, and both nongovernmental groups and government agencies began publishing studies examining how and whether PV could generate a significant part of the nation's electricity.

However, just as these institutions began to make PV a subject of serious analysis and policy consideration, and PV was poised to begin entering the energy mainstream, a set of political dynamics pushed it back out.[14] The most obvious opposition came from fossil fuel industries that opposed renewable energy developments the same way they would any potential competitor. But more important was the particular way in which proponents and opponents characterized renewable energy and its role in society. During the 1970s, groups that advocated for and worked on renewable energy split into deeply hostile factions: conventional and ecological advocates. Conventional advocates, often based in the technical community that worked on engineering research for renewable energy, focused on the technical side of such energy sources and saw them as straightforward substitutes for fossil fuels. In the case of PV, they sought to make the technology cheap enough that one could literally unplug fossil fuel plants and plug renewable energy facilities into their place. They attached no political or social values to their technologies, other than to see them as a technology that could continue the social status quo once fossil fuels became scarce, and do so with much less environmental harm. They were cautious, even skeptical, about the rate at which renewables could diffuse into the broader economy.

In contrast, ecological advocates declared that a society that ran on renewable energy would be a radically different society, more decentralized, ecologically benign, and socially just than the existing society. Opponents of renewable energy agreed with them about a renewable energy society being different, but claimed that the new society would be far from just and prosperous. The rhetoric became quite heated, with renewable energy opponents claiming that the ecological advocates wanted to take society back to a more primitive time while the advocates claiming that they wanted to take it forward to a prosperous but very different future. Those groups debated in front of the public and policy makers, including in Congressional hearings, and in the process the conventional renewable energy advocates got lost in the shuffle, as often in conflict with the ecological advocates as with renewable energy opponents.

These debates, odd as they might seem in retrospect, were very much a part of their time. The energy crisis was only one part of the turmoil of the decade. During the 1970s Americans experienced economic stagnation, the ignominious end of the Vietnam War, the Watergate scandal that drove President Nixon from office, rising crime and divorce rates, increasing and hostile social fragmentation, and the complete dissolution of the elite post-war political consensus. Under these circumstances it is hardly surprising that new energy technologies got politicized; how could they not? Energy, and in particular the rapid increases in fossil fuel consumption, had been deeply embedded in the economic and social development of the United States for centuries.[15] At a time when many social and political institutions seemed to be on the verge of radical changes, it was not too much of a stretch to think that similarly radical changes in energy technologies would also lead to a very different society. For better or worse, a faction of renewable energy advocates and, ironically, conservative political elites both associated renewable energy with a segment of society that had mounted a fierce critique of industrial society. Solar energy symbolized radical environmentalists, not mainstream energy policy. Democratic political elites never accepted that radical vision for society but nonetheless supported renewable energy on more conventional grounds. The more politicized version of renewable energy became manifest in partisan terms when the Republican Ronald Reagan was elected president in 1980 and proceeded to slash government programs that supported renewable energy, along with pushing out Denis Hayes, the director of the government's largest renewable energy research laboratory.

Renewable energy advocates retreated to other callings, pushing for it from other places, including some state governments and in the private sector. By the early 1990s renewable energy was starting to shed its ideological baggage. In 1992, President George H.W. Bush signed into law energy policy legislation that increased government support for renewables and upgraded the government's main renewable energy research laboratory, changing it from a "Research Institute" to a "National Laboratory." However, later in the decade, renewable energy advocates found their technologies politicized again due to debates over global warming. Conservative policy makers who strongly opposed policies to fight global warming associated renewable energy with those policies (as, indeed, did many renewable energy advocates) and the political stalemate over climate change did not help the cause of promoting policies for renewables. A friendly Clinton administration in the 1990s did help keep some supporting policies in place, but pushback from Congress and very low oil prices for most of the decade made energy policy itself a low-salience issue and the resulting policies were short-term and erratic.[16]

By the start of the George W. Bush administration in 2001, it looked as if renewables might start to garner more bipartisan support. The Bush Administration's major energy policy study, completed before 9/11, contained quite a bit of friendly rhetoric for renewables, along with many specific recommendations for policies to support them, policies that had the backing of some Republicans in Congress.[17] The report was not an environmentalist's dream, which is hardly surprising since it came out of a conservative administration headed by men who had worked in the oil industry. In a move that must have galled environmental advocates of renewable energy, the report proposed generating money to support renewables through licensing fees from allowing oil companies to drill in the Arctic National Wildlife Refuge, a project that environmental groups strongly opposed. Nonetheless, renewables enjoyed substantial growth in the first decade of the new century, due in part to federal support and in part to more aggressive support policies in the states, in addition to some large corporate supporters and some high-profile and unexpected pundit support, such as from Thomas Friedman.[18] Renewable energy, at least wind, even picked up some bipartisan support in Congress during Barack

Obama's Administration, as 16 House Republican freshmen, a famously conservative group, asked the House leadership to provide an extension on the production tax credit for wind, the most important federal subsidy for wind.[19] However, the political alignments associated with renewables have not vanished, as evidenced by the rhetoric about them in the 2012 presidential campaign. In the White Paper on energy policy that came out of the Mitt Romney campaign, the only mentions of "renewable" or "solar" energy are in sections that argue for deregulation of the siting of energy facilities.[20] The bankruptcy of Solyndra, a solar firm that had received large loan guarantees from a Department of Energy program, has contributed to this partisan divide, with the company becoming a symbol to conservatives about what is wrong with President Obama's energy policy and his support for renewable energy in particular.[21]

In the context of this politically contentious history, the symbolism of the military's use of renewables, and PV in particular, provides an apolitical, or at least differently political, narrative about solar energy. Whatever one thinks of the U.S. military, it is not a symbol of radical environmentalism or far-left politics and it currently enjoys considerable trust from the public and the apparent respect of political elites of all political persuasion.[22] Images of PV arrays on military bases or of soldiers using small PV systems in the field promote the idea that PV is a serious and important technology, so effective that our military wants it for its facilities and its war-fighting operations, so serious that the lives of our soldiers may depend on it. Which image is more likely to garner bipartisan interest in PV: earnest young people working on a solar bicycle (a bicycle, not a car) or a senior officer in camouflage showing the president of the United States the largest PV array in the country installed in an Air Force base?

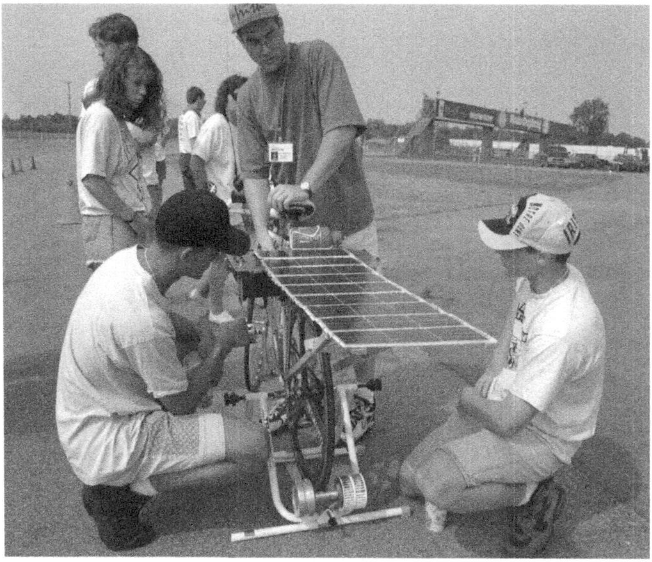

Source: NREL image gallery

The importance of this symbolism has not been lost on firms or advocacy groups involved in PV. Borrego Solar, Kyocera Solar, and Sunpower all feature their military projects on their websites.[23] A search using the term "military" on the website of the Solar Energy Industries Association (http://www.seia.org) brings up 20 press stories and releases about assorted military solar projects or statements. The website of the American Solar Energy Society has, among other things, a story about Dorothy Robyn, former Deputy Under Secretary of Defense for Installations and Environment, U.S. Department of Defense that includes a discussion of how important DOD work is to solar energy. Given that DOD represents roughly 0.5% of the PV market, the size of military purchases is not a sufficient explanation for featuring them so prominently.

Source: Nellis Air Force Base photos

In short, the military adoption of renewable energy, and PV in particular, can go a long way toward remaking the symbolic profile of the technology. It may help to dispel the industry's problem that Jeff Himmelman so aptly described in the August 12, 2012 Sunday *New York Times Magazine* article: "The first [problem the industry has] is abstract and well ingrained in the American psyche: the negative association of 'green' technologies with inefficiency and idealistic, hippie-fueled impracticality."[24] The image of soldiers using PV in combat will help to refute that idea better than any economic analysis.

How do symbolic orders change? How does the meaning people attach to an object or an inscription change? This is not the sort of change that one can simply will to happen or create with a simple stroke. In the case of PV, and renewable energy technologies more generally, different groups have contested the symbolic meaning attributed to them, but at different times one meaning or another has been dominant in elite policy-making cultures. Part of what leads to symbolic change is the use that groups make of the technology and the manner in which they present those uses. In this sense, the DOD research on and procurement of PV technologies is a signal part of changing their symbolic meaning, since a military imprimatur on the technology goes a long way to making it one that is important in American society. In that sense, PV firms and advocacy groups are doing just what they should be doing, publicizing how important these technologies are to the military and putting photographs of military use anywhere they can.

That said, conscious efforts to impart a different symbolic meaning are only part of what drives such change. The other part is the historically specific social context in which such changes take place. To what extent will the DOD come to view PVs as truly important to its strategic mission? To what extent are legislators, bureaucrats, businesspeople, and the general public ready to embrace a substantial change in their energy system? What kind of change, socially, culturally, and economically, does PV represent? The answers to these questions will emerge from unpredictable combinations of factors, and DOD's use of PV will only be one part of this equation.

Policy Recommendations

One easy recommendation to make is that DOD continue to seek, and if possible expand, opportunities to employ PV in the service of the agency's security mission. The potential for DOD action, in both procurement and R&D, is immense and current programs have only scratched the surface. But such expansion is no sure thing, and DOD use of PV could contract. Much of that procurement is driven by investment tax credits for PV, which expire in 2016, unless Congress extends them further,[25] leaving the industry in a potentially bad place after that time. In addition, the history of investment tax credits for solar energy in the 1970s and 80s is a grim reminder that this particular kind of subsidy can lead to industry problems. Investment tax credits to not reward producing electricity so much as getting equipment in the ground, and poorly maintained or low-quality installations will damage the industry's reputation as whole.

One lesson from other technologies that DOD has procured is to keep close tabs on the particular firms that sell to DOD. Firms that are almost exclusively defense contractors have a poor reputation for efficiency or the ability to sell into a civilian market, where price drives decisions more than performance. Firms that sell both to DOD and the civilian market keep their competitive edge and simply have another buyer available to them. DOD will help to catalyze the growth of the PV market only to the extent that it patronizes firms that can also work in the civilian market. Large DOD installations can partly avoid this problem using the model that several bases have already used. Those bases do not buy the solar systems but rather contract with a third party to buy and install them. DOD procures solar electricity, not the solar panels themselves, and the third parties that build the systems can shop for the best deals on the hardware itself. Still, the arcana of DOD procurement could provide opportunities for too-cozy relationships between DOD and solar suppliers, and the field would benefit from several case studies of firms that have been successful—or not—in selling into both military and civilian markets.

---

[1] U.S. Energy Information Administration, *Solar Photovoltaic Cell/Module Shipments Report 2010* (Washington, DC: U.S. EIA, Jan. 2012), available at: http://www.eia.gov/renewable/annual/solar_photo/archive/solarpv10.pdf

[2] Charles Weiss and William B. Bonvillian, *Structuring an Energy Technology Revolution* (Cambridge, MA: The MIT Press, 2009).

[3] John Perlin, *From Space to Earth: The Story of Solar Electricity* (Ann Arbor, MI: Aatec Publications, 1999); Frank N. Laird, *Solar Energy, Technology Policy, and Institutional Values* (New York, NY: Cambridge University Press, 2001).

[4] Their table includes concentrating solar power installations, which usually do not mean PV, so those facilities have been excluded from Table 1 here.

[5] "Major Solar Projects List," Solar Energy Industries Association website (19 Aug. 2014), available at: http://www.seia.org/research-resources/major-solar-projects-list

[6] U.S. Department of Energy, "U.S. Marine Corps Base Camp Pendleton: Using the Sun for Hot Water and Electricity," summary paper, U.S. DOE, Energy Efficiency and Renewable Energy, Federal Energy Management Program (Sept. 2009), available at: http://www1.eere.energy.gov/femp/pdfs/46348.pdf; U.S. Environmental Protection Agency, "Success stories—Siting Renewable Energy on Contaminated Land: Fort Carson, Colorado," U.S. EPA factsheet (Feb. 2009), available at: http://www.epa.gov/oswercpa/docs/success_fortcarson_co.pdf; and U.S. Environmental Protection Agency, "Success stories—Siting Renewable Energy on Contaminated Land: Nellis Air Force Base Nevada, "U.S. EPA factsheet (Feb. 2009), available at: http://www.epa.gov/oswercpa/docs/success_nellis_nv.pdf

[7] "U.S. Solar Market Insight Report: 2013 Year in Review," Solar Energy Industries Association website, available at: http://www.seia.org/research-resources/solar-market-insight-report-2013-year-review

[8] "U.S. Solar Market Insight Report: 2011 Year-in-Review Executive Summary," Solar Energy Industries Association (2012), available at:
http://www.seia.org/sites/default/files/resources/2011%20Q4%20SMI%20ES.pdf

[9] U.S. Government Accountability Office, *Renewable Energy: Federal Agencies Implement Hundreds of Initiatives* (Washington, DC: U.S. GAO Report to the Committee on Homeland Security and Governmental Affairs, U.S. Senate, GAO-12-260 (Feb. 2012).

[10] U.S. Energy Information Administration, *Solar Photovoltaic Cell/Module Shipments Report.*

[11] U.S. Department of Energy, *Department of Energy FY 2015 Congressional Budget Request,* Vol. 3 (Washington, DC: U.S. DOE, Office of the Chief Financial Officer, DOE/CF-0098, Mar. 2014).

[12] Perlin, *From Space to Earth,* ch. 3 and 5.

[13] Laird, *Solar Energy.*

[14] Frank N. Laird, "Constructing the Future: Advocating Energy Technologies in the Cold War," *Technology and Culture* 44, no. 1 (Jan. 2003): pp. 27-49.

[15] David E. Nye, *Consuming Power: A Social History of American Energies* (Cambridge, MA: MIT Press, 1998).

[16] Peter S. Fox-Penner, "Instruments and Tools in Clinton-Era Policy," ch. 17 in David Lewis Feldman, ed., *The Energy Crisis: Unresolved Issues and Enduring Legacies* (Baltimore, MD: Johns Hopkins University Press, 1996): pp. 256-263; Paul Joskow, *U.S. Energy Policy During the 1990s* (Cambridge, MA: National Bureau of Economic Research Working Paper No. 8454, Sept. 2001).

[17] National Energy Policy Development Group, *National Energy Policy: Reliable, Affordable, and Environmentally Sound Energy for America's Future* (Washington, DC: U.S. Government Printing Office, 2001).

[18] Galen Barbose, Naïm Darghouth, Ryan Wiser, and Joachim Seel, *Tracking the Sun IV: An Historical Summary of the Installed Cost of Photovoltaics in the United States from 1998 to 2010* (Berkeley, CA: Lawrence Berkeley National Laboratory Report LBNL-

5047E, Sept. 2011); Ryan Wiser and Mark Bolinger, *2011 Wind Technologies Market Report* (Oak Ridge, TN: U.S. DOE, Report DOE/GO-102012-3472, Aug. 2012); Thomas L. Friedman, "The New Red, White, and Blue," *New York Times* (6 Jan. 2006): p. 21.

[19] "16 Republican House freshmen petition leadership for immediate PTC extension," American Wind Energy Association press release (27 June 2012).

[20] Mitt Romney and Paul Ryan, "The Romney Plan for a Stronger Middle Class: Energy Independence," campaign document (23 Aug. 2012).

[21] Amy Harder, "Deconstructing Solyndra," *National Journal* (22 Nov. 2011).

[22] Lydia Saad, "Congress Ranks Last in Confidence in Institutions," Gallup Politics website, Gallup Poll Briefing 1 (22 July 2010), available at: http://www.gallup.com/poll/141512/congress-ranks-last-confidence-institutions.aspx

[23] "Solar Energy Projects," Borrego Solar website (2014), available at: http://borregosolar.com/solar-projects/all/; "Government," Sunpower website (2014), available at: http://us.sunpower.com/commercial-solar/local-state-federal-government/

[24] Jeff Himmelman, "The Secret to Solar Power," *New York Times Magazine* (9 Aug. 2012), p. MM24.

[25] See the federal incentives section on the Database on State Incentives for Renewables and Efficiency, available at: http://www.dsireusa.org/incentives/index.cfm?state=us

5

THE GOVERNMENT ROLE IN DEVELOPING SOLAR THERMAL TECHNOLOGY

Miles Brundage

Introduction

Solar thermal power, also known as concentrating solar power, is a renewable source of energy that captures solar radiation the form of heat and uses that heat to drive a turbine that generates electricity. (Solar photovoltaic cells, in contrast, directly convert sunlight to electricity). Several large-scale solar thermal power plants have been built in the United States, and more are currently under construction in the Southwestern United States, California, and in countries including Spain, Israel, South Africa, and India. Compared to solar photovoltaics, solar thermal energy has the advantage of being able to store energy for hours in the form of heat, which allows production of electricity into the night and smoothing out of fluctuations when the weather is cloudy. This storage capability, as well as the relative technical simplicity of solar thermal systems, and the availability and reasonable cost of constituent materials and components, has led to rapid growth in the market for solar thermal power plants, and promises substantial potential for future growth.

The U.S. federal government has played a significant role in the development of solar thermal technology over the past four decades. This influence has occurred through a variety of mechanisms, including funding of basic research at universities and research and development at National Laboratories, conducting demonstration projects in partnership with industry, implementing subsidies, and developing solar radiation maps for the United States to aid industry in siting power plants.

Solar Thermal Power Fundamentals

There are four main types of solar thermal power plants, characterized by the different ways that they concentrate solar radiation before using the collected heat to generate electricity:

- Parabolic trough systems use a reflective trough to concentrate sunlight on a receiver tube running parallel to the trough.
- Linear Fresnel systems use an array of flat or slightly curved mirrors to concentrate sunlight on a downward-facing linear receiver.
- Power tower systems use an array of heliostats (tracking mirrors) to focus sunlight on a central tower where the generator is located.
- Dish-engine (or dish-Stirling) systems use a series of parabolic dishes to concentrate sunlight on an engine located directly above the dishes (rather than at a central location at the power plant, as in the other system designs).[1]

These four varieties of solar thermal power generation have different levels of technological maturity, and each has its own advantages and disadvantages in terms of water requirements, land area requirements, and maximum operating temperature (which is correlated with the efficiency of the system).[2] Parabolic trough systems are the most mature and make up the majority of the solar thermal systems currently in operation. Linear Fresnel systems are relatively new and untested, and have primarily been developed outside the United States, but have some possible advantages such as the relative ease of manufacturing flat reflectors compared to curved ones. Power tower systems have the potential to operate at higher temperatures and efficiencies than other approaches and as such are currently receiving significant industry attention, though they still lag behind parabolic trough systems in maturity and market penetration. Finally, dish-engine systems have been proven to be feasible since the 1980s and have some advantages such as low water requirements and flexibility to work in either a distributed or utility-scale fashion (since each dish independently provides electricity rather than requiring a central generator for a whole plant, as with power tower systems). However, dish-engine systems are currently unable to store energy, though some research is being done to address this limitation.

History of Solar Thermal Technology Development

In response to oil crises of the 1970s, the U.S. federal government began investing millions of dollars annually in the development of solar thermal technology along with other renewable technologies such as photovoltaics and wind power.[3] Support for solar thermal began through the National Science Foundation with research on power towers, but since those early efforts, most federal support has come from the U.S. Department of Energy (DOE) and its network of National Laboratories.[4] In addition to research and development support, financial incentives for solar thermal power plants have played a critical role in the commercialization of solar thermal technology, such as through tax incentives administered by the U.S. Department of the Treasury and loan guarantees from DOE.

Figure 1. DOE Funding for CSP Technologies

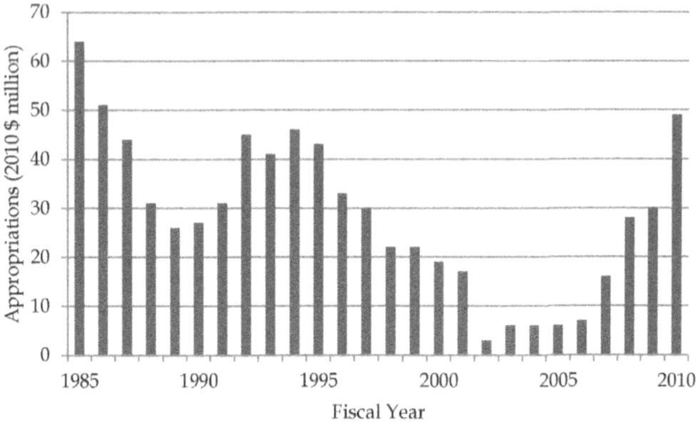

Source: Recreated from the U.S. Department of Energy, *Report on the First Quadrennial Technology Review: Technology Assessments* (August 2012).

The DOE has funded solar thermal research and development in industry and academia since the late 1970s and has conducted its own research through the National Laboratories. Laboratories such as Lawrence Livermore National Laboratory and Sandia National Laboratories, which have historically been focused on weapons research, have been major sites of solar thermal research over the past three decades in response to the 1970s' oil crises and the characterization of energy supply as a national security issue by national leaders beginning with President Carter. Sandia, for example, established a solar thermal testing facility in 1979 which is still used by industry partners to test new solar thermal collectors, generator, and thermal storage designs under controlled conditions. The Solar Energy Research Institute, or SERI (later renamed the National Renewable Energy Laboratory, or NREL) has also been prominently engaged in research on both photovoltaic and solar thermal power systems, and leads the solar resource assessments which industry draws on to predict power plant output and reliability for a given geographic area.[5]

The first commercial solar thermal power plant in the United States was Solar Energy Generating Systems (SEGS), which was built in stages (SEGS I through SEGS IX) from 1983 to 1990. Located in California's Mojave Desert, SEGS remains in operation today with over 350 MW of capacity.[6] SEGS was built by Luz Industries and helped to demonstrate the feasibility and reduce the costs associated with parabolic trough solar thermal systems. Luz ultimately went bankrupt, citing fluctuation in government policies as one of the reasons for its unprofitability, but investors in particular facilities (which Luz raised capital for individually) were able to recoup their investments.[7] While the plant was built by a private company with no direct funding from DOE, Luz benefited from prior federal research in parabolic troughs and specialized coatings for pipe-receivers. Additionally, Luz was able to invest in a technology that was untested at the time partially due to government incentives, most notably a 10% investment tax credit (ITC) and a 15% energy tax credit (ETC) instituted by the Energy Tax Act of 1978, as well as a state-level tax credit. Lastly, the Public Utilities Regulatory Policy Act (PURPA) of 1978 influenced Luz's investment decisions by requiring utilities to purchase electricity from generating facilities that meet certain requirements, thus incentivizing Luz to build plants meeting these requirements.[8] As an example of the influence of this legislation, PURPA affected Luz's decisions on the sizing of individual facilities—specifically, plants were designed so as to avoid exceeding 30 MW and 80 MW cut-off levels for eligibility, so several SEGS sites have precisely those power levels. While a Luz executive later complained that such size cut-offs were harmful by preventing more flexible engineering of the plants (solar thermal plants have different ideal sizes based on land availability, design, storage, etc.), PURPA's guarantee of a market seems to have been essential to Luz's temporary success with SEGS.

During the 1980s and 1990s, two pilot projects in power tower technology—Solar One and Solar Two—were developed by a consortium of government and industry contributors (including DOE, Southern California Edison, and the California Energy Commission), with the federal government contributing half of the cost.[9] The Solar One and Solar Two projects accelerated technical progress in understanding the challenges of operating such plants, which were considered too financially risky to be developed solely by private investors at the time.[10] Sandia National Laboratory and NREL led the design and testing of these projects. Based in part on these demonstrations that large-scale power towers were feasible, a commercial power tower plant has since been built in California and many are under construction around the world. Industry has benefited from lessons learned through the projects;[11] in particular, the Solar Two plant helped validate molten salt as a means of energy storage, which has since been adopted in other facilities.[12] A key advantage of power towers comes from the high operating temperature they produce, which leads to higher efficiency and lower thermal storage costs. Their high temperatures also make power towers suitable for incorporation in hybrid fossil fuel-solar plants, and power tower systems are scalable to a wide range of plant sizes.[13]

The federal government also played a critical role in the development of dish-engine technologies through early research in the 1970s and 1980s at NASA's Jet Propulsion Laboratory and the DOE weapons labs, and through later collaborations with industry. In the early 1990s, Sandia National Laboratories and NREL partnered with Cummins Power Generation on a 50-50 cost-shared joint venture to develop dish-engine systems, though Cummins ultimately abandoned the technology.[14] While DOE continues to fund research on improving dish-engine technology,[15] one of the main companies working on such technologies, Stirling Energy Systems, filed for bankruptcy in 2011, and the future of the technology is unclear, with only one operating facility at present and one under construction.[16]

The federal government has provided various financial incentives to energy producers over the last several decades, including some that benefitted solar thermal technology. As noted earlier, construction of the first (and to this day, still the largest) solar thermal plant, SEGS, was incentivized through federal tax credits instituted in 1978, but fluctuation and uncertainty in these incentives created difficulties for making long-term investments.[17] Historically, tax credits for renewable energy in the United States have been extended near the time of their expiration, but some lapses have occurred, and lack of long-term predictability may deter corporate commitments to these technologies. Currently, the relevant investment tax credit for solar is set to last through 2016.[18] Other financial incentives such as accelerated depreciation of capital through the Modified Accelerated Cost Recovery System (MARCS), established in 1986, have also benefited the solar thermal industry.[19]

The American Recovery and Reinvestment Act of 2009 (the "Recovery Act" for short, also known as the stimulus package) invested billions of dollars in scaling up solar energy production, including solar thermal, through direct grants as well as loan guarantees. This has led to several large-scale solar thermal projects that are currently under construction or in operation, including some of the largest solar thermal plants ever built.[20] The DOE Loan

Guarantee Program has supported projects such as Abengoa's 280 MW parabolic trough plant near Gila Bend, Arizona; NextEra Energy Resource's 250 MW parabolic trough plant in Riverside, California; and NRG Energy and Brightsource Energy's 377 MW power tower project in Ivanpah, California (note that Brightsource Energy was founded by many of the executives involved in Luz's construction of SEGS, discussed earlier).[21] The Treasury Department's 1603 Grant Program, which has since been allowed to expire, gave grants to companies in lieu of tax credits and supported the one operating dish-engine system, in Peoria, Arizona.[22] Of these, the Solana plant by Abengoa in Gila Bend, Arizona, has received particular attention both as the first commercial scale plant with molten salt storage (a technology which was demonstrated by the DOE in the 1990s) and as the biggest parabolic trough plant in the world.

Another recent development in the federal government's involvement in solar thermal technology is the launch of DOE's SunShot Initiative, which is focused on reducing the cost of solar energy (solar thermal and photovoltaic) substantially by the end of the decade to make it cost-competitive with fossil-fuel energy plants without the need for subsidies. The SunShot Initiative's "Vision Study" concluded that a combination of evolutionary and revolutionary improvements to solar photovoltaic and solar thermal technologies will be needed to bring their costs down to the agency's targets.[23] SunShot has recently funded a portfolio of research projects in academia, industry, and National Laboratories aimed at goals such as increasing system efficiency through higher temperature operations, reducing optical and thermal efficiency losses, and reducing the overall costs of building solar plants, including efforts aimed at reducing the "soft costs" (such as land acquisition and preparation, construction labor, and maintenance) of both solar photovoltaic and solar thermal technologies.[24]

Status and Prospects for Solar Thermal

Solar thermal power has advanced greatly over the past few decades, with notable improvements supported by DOE including the validation of power tower systems, improvements to the efficiency of dish-engine systems, and demonstration of feasible means of energy storage, which is critical in enabling solar thermal to compete directly with fossil fuel sources of electricity in the utility-scale power generation market and improving the economics of large-scale plants. The SunShot Vision Study outlining DOE's view of the possible future of solar technologies identifies many opportunities for further cost reduction which, if achieved, potentially allow solar power (both solar thermal and photovoltaics) to provide more than ten percent of U.S. electricity demand by 2030 and more than a quarter by 2050.[25] While past projections of the growth of solar energy production have typically been over-optimistic, the industry has, indeed, grown substantially in the U.S. in the last few years, in part due to Recovery Act funding, with installed capacity more than doubling since 2008—moreover, dozens of plants are currently under construction or recently announced worldwide, with 410 MW of solar thermal capacity installed in 2013 alone.[26]

Several factors place practical constraints on the current and future growth of solar thermal technology in the United States, including: water requirements (depending on the specific technology used); the need for financing for first-of-a-kind technologies such as new power tower, Fresnel lens, and dish-engine systems; legal and logistical issues involved in obtaining and preparing land for development; and the fact that the highest quality solar resources are generally in areas that are sparsely populated, so transmission and distribution infrastructure for the produced electricity must also be developed. On the other hand, solar thermal technologies have potential economic advantages in that the supply chain is largely domestic (unlike, for example, photovoltaics, where much technical and production capacity lies overseas) because solar thermal systems use many components that are commonly used in other technologies. Energy storage continues to differentiate solar thermal from photovoltaic (though photovoltaic has other advantages, including no water requirement during operation and more flexibility in terms of plant size), and molten salt storage is now well-established and can greatly improve both generating capacity and overall project economics associated with utility-scale solar thermal plants. To continue to leverage these potential advantages, DOE's SunShot Initiative is currently investing tens of millions of dollars per year in research to further reduce the cost of the solar thermal technology.

Lessons for Technology Policy

Several lessons emerge from examining the history of U.S. involvement in solar thermal technology. First, stability of incentives and research and development efforts over time is critical for technology development. DOE's level of support for renewable energy research and technology innovation has fluctuated wildly since the late 1970s, which has had both direct and indirect negative effects on technology development. Solar thermal has gone through periods of boom and bust, driven in part by the availability or scarcity of government incentives, and uncertainty about the future of tax incentives. Such fluctuations negatively impacted the development of the technology both by creating market uncertainty and by dampening morale in DOE and making it difficult to attract talented program managers.[27] Another key lesson is the crucial role of government in supporting a diversity of early-stage technical approaches to a given problem. The four solar thermal technologies discussed here have all been supported to varying extents by the U.S. federal government, since each had (and still has) both significant uncertainties, and potential benefits, in terms of future performance and costs. Had power tower research been abandoned due to the predominance of parabolic trough systems in the market, then the renaissance in that technology today, with plants such as that at Ivanpah set to generate hundreds of megawatts of power, might not have occurred. At the same time, technologies supported by the government such as dish-engine systems have not always caught on as hoped, and it is unclear to what extent that research will ever pay off in a major way. Yet innovation paths are often highly uncertain, especially in the early stages of a technology's evolution, and a crucial government function can be to help keep a variety of technological alternatives open as markets explore the benefits of competing pathways. To this day, the future mix of power tower versus parabolic trough and other types of solar thermal plants is unclear, and the DOE remains engaged in all these areas. In the case of solar thermal, the federal government has used an array of tools, often in close partnership with the private sector, ranging from university-based R&D, to technology demonstration projects, to various investment incentives, aimed at improving the technological base, and the capacity for learning in the private sector.

---

[1] U.S. Department of Energy, *Report on the First Quadrennial Technology Review: Technology Assessments* (Washington, DC: U.S. DOE, Aug. 2012).

[2] Ibid.

[3] Donald Beattie, ed., *History and Overview of Solar Heat Technologies* (Cambridge, MA: MIT Press, 1997).

[4] U.S. Department of Energy, *First Quadrennial Technology Review.*

[5] "Renewable Resource Data Center," National Renewable Energy Laboratory website, available at: http://www.nrel.gov/rredc/solar_data.html

[6] "Energy Timelines: Solar Thermal," U.S. Energy Information Administration website, available at: http://www.eia.gov/kids/energy.cfm?page=tl_solarthermal

[7] Michael Lotker, *Barriers to Commercialization of Large-Scale Solar Electricity: Lessons Learned from the Luz Experience* (Albuquerque, NM: Sandia National Laboratories Contractor Report, SAND91-7014, 1991).

[8] Ibid.

[9] U.S. Government Accountability Office, *Electricity Supply: Efforts Under Way to Develop Solar and Wind Energy* (Washington, DC: Subcommittee on Investigations and Oversight, Committee on Science, Space, and Technology, House of Representatives, 1993).

[10] W.R. Gates, *Solar Thermal Technology Development: Estimated Market Size and Energy Cost Savings* (Pasadena, CA: Jet Propulsion Laboratory, National Aeronautics and Space Administration, 1983).

[11] "List of solar thermal power stations," Wikipedia website, available at: http://en.wikipedia.org/wiki/List_of_solar_thermal_power_stations

[12] Tomislav M. Pavlović, Ivana S. Radonjić, Dragana D. Milosavljević, and Lana S. Pantić, "A review of concentrating solar power plants in the world and their potential use in Serbia," *Renewable and Sustainable Energy Reviews* 16 (2012): pp. 3891-3902.

[13] U.S. Department of Energy, *First Quadrennial Technology Review.*

[14] U.S. Government Accountability Office, *Electricity Supply.*

[15] "Dish Engine," U.S. Department of Energy SunShot Initiative website, available at: http://www1.eere.energy.gov/solar/sunshot/csp_dish.html

[16] Pavlović, et al., "Review of concentrating solar power plants."

[17] Lotker, *Barriers to Commercialization.*

[18] "Solar Investment Tax Credit (ITC)," Solar Energy Industries Association website, available at: http://www.seia.org/policy/finance-tax/solar-investment-tax-credit

[19] "Depreciation of Solar Energy Property in MACRS," Solar Energy Industries Association website, available at: http://www.seia.org/policy/finance-tax/depreciation-solar-energy-property-macrs

[20] "Our Projects," U.S. Department of Energy Loan Programs Office website, available at: https://lpo.energy.gov/?page_id=45

[21] Ibid.

[22] "Top 10 Solar 1603 Treasury Grant Awards," Gunther Portfolio website (10 Dec. 2010), available at: http://guntherportfolio.com/2010/12/top-10-solar-1603-treasury-grant-awards/

[23] U.S. Department of Energy, *SunShot Vision Study* (Washington, DC: U.S. DOE, National Renewable Energy Laboratory, DOE/GO-102012-3037, Feb. 2012).

[24] "Concentrating Solar Power SunShot Research and Development," U.S. Department of Energy SunShot Initiative website, available at: http://www1.eere.energy.gov/solar/sunshot/csp_sunshotrnd.html

[25] U.S. Department of Energy, *SunShot Vision Study.*

[26] "List of solar thermal power stations," Wikipedia; "New Report: U.S. Solar Market Grows 41%, Has Record Year in 2013," Solar Energy Industries Association website (4 Mar. 2014), available at: http://www.seia.org/news/new-report-us-solar-market-grows-41-has-record-year-2013

[27] Donald Beattie, ed., *History and Overview of Solar Heat Technologies* (Cambridge, MA: MIT Press, 1997).

6

FEDERAL INVESTMENTS IN ADVANCING SOLAR PHOTOVOLTAICS

Jason O'Leary

Introduction

Federal government investments in solar photovoltaic (PV) research and development (R&D) over the past six decades can be divided into four eras. During the 1950s and 1960s the funding for solar PV innovation came from military-industrial partnerships and satellite contracts for the space race. The 1970s was a time of transition, as national attention shifted toward environmental and resource concerns, and government support for solar PV research shifted to terrestrial applications of solar PV and establishing institutions to support that endeavor. The 1980s began a long era of diminished funding for R&D, and an increased focus on market-oriented solutions, including government partnerships with industry and academia, and heightened incentives for technology transfer. Starting in 2006, a new era of solar PV investment arose in light of renewed concerns around global economic, social, and environmental instability. R&D funding increased, as did support for start-up firms in a period of shrinking venture capital. Nonetheless, as a part of the overall energy innovation landscape, federal investments in solar R&D remain just a small fraction of the vast R&D funding levels for nuclear and coal.

Sixty years of federal investment in solar R&D are also characterized by the evolution and increasing diversity of strategies pursued by the government. Yet these policy strategies have also become less direct, moving from government acting as the primary investor and user in the first era, to a more complex, supporting role in later eras. This chapter thus discusses a range of federal activities, including providing public information, offering loan guarantees and incentives, forming and maintaining repositories of technical information, encouraging technology transfer, and of course, supporting R&D, which together constitute a complex and comprehensive approach to advancing solar PV innovation.

Summary of Solar PV Technology

"Solar cells" are the basic building blocks of solar PV systems. "Solar panels," also called "solar modules," consist of a group of solar cells linked together as one unit. Multiple solar panels are linked together to form a system installation, or an "array." Solar PV installations may range anywhere from 4-5 kilowatts for an average residential rooftop system, up to several hundred megawatts for a large utility-scale system.

Basic Principles

Most solar PV cells are made up of layers of silicon and other materials for various purposes such as charging, reflection or protection. Two layers of silicon are coated or "doped" with substances that give one side a positive or "p" charge, and the other side a negative or "n" charge.

When sandwiched together, these layers create a "p-n junction." The p-n junction creates an electric field, which is like a one-way street for electrons to flow from the n side to the p side. When photons in sunlight hit the silicon, electrons break free to travel across this p-n junction, thus creating an electrical current.[1] The electricity is captured by metal contacts and then flows in the form of direct current to an inverter, which turns it into our standard alternating current.

Types of Solar Cell Technology

Solar PV cells are commonly sorted into three generations, classified by their time of research origin, as well as component materials. First generation cells use silicon and are fairly efficient and reliable. The evolution of first generation silicon solar PV cells has benefitted from R&D in the semiconductor field, which also depends on silicon. Indeed, silicon PV cells have maintained their dominance in part through synergies with semiconductor R&D, and as well because the scale of the semiconductor industry has ensured continued availability of, and a strong market for, silicon. Nonetheless, until recent silicon price wars, silicon cells remained expensive compared to later-generation technologies.

Second generation cells, known as "thin-film" cells, are generally less efficient than silicon cells, but, until recently, they were also less expensive due to the manufacturing process and the types of materials used. Indeed, replacing silicon with other low-cost, readily available materials had been one of the main ambitions of R&D. Companies such as Eastman Kodak and Boeing, as well as public institutions such as University of South Florida and the Department of Energy's National Renewable Energy Laboratory (NREL), all worked on versions of thin-films that included copper and cadmium telluride (CdTe), which became the dominant technology on the thin-film market in the early 2000s, when First Solar Inc. and several other companies began using this compound to manufacture solar PV on a large scale.

Third generation cells represent the current cutting-edge of the field, making use of advanced materials and processes with a focus both on cost-competitive, streamlined manufacturing and sustainable manufacturing practices. The most established of the third generation approaches are "multijunction" and "concentrator" technologies, and these are often combined together into a single system. Multijunction cells use several layers to capture more light at different wavelengths. Concentrator systems use lenses, mirrors, or reflectors to concentrate the sunlight to many times its normal intensity onto a small area of highly efficient solar PV cells. By using fewer expensive solar PV cells (such as multijunction) and more inexpensive materials such as lenses and reflective surfaces, the overall cost of a concentrator system is comparable to systems using lower-efficiency cells. However, the mechanical requirements are more complex, increasing operations and maintenance concerns.

Table 1. Solar PV Technology at a Glance

	Technology	Age (yrs.)	Current Eff. Ratings	Avg. Mod. Price (2011 U.S. $)	Primary Country of Origin
Gen 1	Silicon	55	20-27%	$1.63	China, Philippines
Gen 2	Thin-film	35	13-20%	$1.28	Malaysia
Gen 3	Multijunction, Concentrator	8-30	26-44%	$1.32	United States

History of Solar Photovoltaic Innovation: Four Eras

Since the 1950s, approaches to innovation in solar PV correspond to changes in society and in beliefs about the appropriate role for government in technology development. As the environmental movement peaked, Cold War tensions rose and fell, and views about the role of government in society changed, so did the funding priorities and institutional models for solar PV and other energy technologies. The funding mechanisms ranged from the issuance of defense contracts to private firms, to the creation of national institutions for energy R&D, to demand-side and market-based economic approaches. Public-private partnerships played a role throughout, though especially after 1980 when the Bayh-Dole Act provided new incentives for academic scientists to seek patents for their inventions and discoveries.

Figure 1. DOE R&D Appropriations for Solar PV, 1978-2013

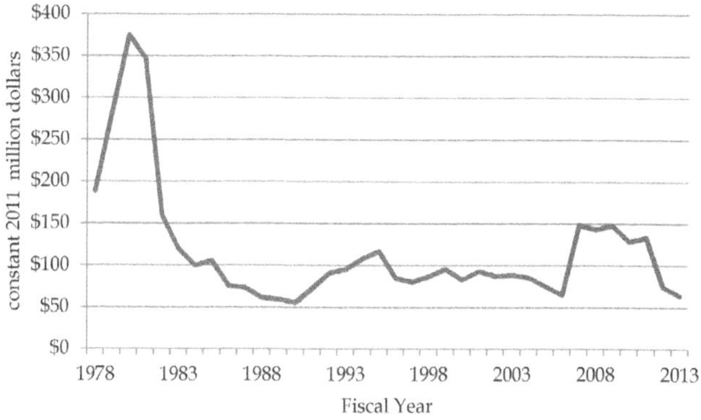

Source: EIA detailed budget reports.

Over time, a narrow focus on R&D contracting broadened to include tax incentives that stimulated market demand, and loan guarantees that supported businesses.

While the federal government funded R&D through this entire period, level of investment has varied widely, peaking in the late 1970s and early 1980s, and reaching their nadir between 1982 and 2006. These changing commitments can be divided into four distinct eras.

Era I: Military Contracts to Civilian Energy, 1955-1970

The first era of solar PV technology was mostly funded by government contract to private companies that performed the research and development. Spurred on especially by the Soviet Union's Sputnik satellite launch in 1957, this early work was particularly focused on applications for spacecraft, and is best understood as just one small element of the broader Cold War competition. The dynamics of the public-private approach pursued during the early days of solar PV R&D in the 1950s is well illustrated by the efforts of two entrepreneurs, Alfred Mann and Leslie Hoffman. Mann was a physicist working for Technicolor when he started his own company, Spectrolab, in 1956 to conduct work for the Army on missile guidance systems. The Air Force contracted with Spectrolab two years later to build solar cells for their spacecraft.[2] Among other advances, Spectrolab created the coating to help the solar cells remain cooler, and thus more efficient when deployed in space.

Hoffman Electronics was Mann's main competitor in the Los Angeles area in the 1950s, producing the solar PV modules for Vanguard I, the first satellite to use PV technology. Hoffman Electronics made major improvements to solar cells in those early years, bringing the efficiency from 8% in 1958 to 14% in 1960. While the market was new and small by today's standards, these two leaders in the field vied for engineering talent and contracts with the Department of Defense (DOD) and the National Aeronautics and Space Administration (NASA), thus helping to start the world's first solar PV business cluster in the

Los Angeles area.[3] Spectrolab, now a subsidiary of Boeing, continues to develop some of the most efficient solar PV cells.[4]

Era II: Major Energy Transitions, 1970-1982

The second era of solar PV innovation focused on the establishment of large research initiatives, in the context of a government rationale for supporting solar PV R&D that moved far beyond the narrow application for space technology to the broader quest for providing new energy sources for society. The main stimulus for this transition was of course the oil crisis of 1973, as well as the environmental movement. The policy approach shifted from narrowly focused funding of space-race contracts to private-sector firms, to broad support for government agencies and other research institutions with the goal of accelerating terrestrial solar PV innovation. This new priority was first signaled in 1973, when the NSF sponsored a workshop in Cherry Hill, New Jersey called "Photovoltaic Conversion of Solar Energy for Terrestrial Applications," which brought together government, academia, and private industry representatives to coordinate research and development efforts and standards for Solar PV. [5]

With the new rationale for government support of solar PV research, and increased levels of investment, came new institutional settings for R&D. In 1974, President Carter signed the Solar Energy Research Development and Demonstration Act, which created the Solar Energy Research Institute (SERI). Three years later the Department of Energy (DOE) was created. The government had become both a major sponsor and performer of solar PV innovation, with R&D funding between 1978 and 1982 totaling nearly $1.4 billion, and annual expenditures reaching a high of $375 million in 1980.[6] While this was still only a fraction of the appropriations for nuclear R&D ($4.9 billion), it nonetheless represented an enormous increase in effort from the funding levels of the 1960s, which were in the hundreds of thousands of dollars (precise spending levels cannot be determined as agencies did not report specific solar funding program expenditures prior to the mid-1970s).[7]

NASA also began to explore terrestrial applications for solar PV. In 1975 NASA's Jet Propulsion Laboratory led the effort and helped coordinate the birth of the Flat Plate Solar Array (FSA) project.[8] This project started with crystalline-silicon, a proven material used in solar PV cells for spacecraft during previous decades, and tasked researchers with improving the technology for earth-based applications. Efficiency, durability, and cost were major concerns.[9] Unlike in the vacuum of space, terrestrial-bound solar panels had to contend with environmental degradation in the earth's atmosphere. Under this research program, the solar cell lifespan increased from 1 year in 1975 to 10 years by 1986. Meanwhile, DOE-funded R&D included pilot projects for grid-tied, residential rooftop solar PV installations as well as large solar arrays utilized for crop irrigation and drying.[10] These government programs signaled the potential for expanded civilian markets and helped to attract private sector involvement. A notable success was Exxon's support for independent scientist Elliot Berman, whose work helped to reduce the cost of solar cells 80% by using a lower grade of silicon and less expensive materials. Research conducted at Union Carbide and funded by FSA and the DOE yielded a new generation of manufacturing processes and automated machines that helped produce solar cells more quickly, efficiently, and with a higher level of quality. R&D sponsored by Boeing and Kodak between 1979 and 1982[11] led to improved efficiency of thin-film technologies.[12]

Era III: Market-Oriented Policies and Support for the Innovation System, 1982-2006

The third era of solar PV innovation was characterized by market-based solutions and overall reduced funding for R&D. While existing institutions and agencies continued to conduct research, they did so at a fraction of their previous federal funding levels, while a wider variety of policy tools were deployed to advance PV, such as public education, facilitating technology transfer, and setting performance standards.

The solar energy enthusiasm of the 1970s was quickly brought to an end by the Reagan Administration, which cut energy R&D overall and continued to favor nuclear and coal over solar. Funding for solar PV R&D was cut from $375 million in 1980 to approximately $100 million for fiscal year 1984.[13] While considerably reduced in scope, SERI continued to be the main locus of government R&D on solar PV until 1991, when President Bush oversaw its transformation into the National Renewable Energy Lab (NREL), an organization with a much broader and complex mission than its predecessor. Not only did NREL become the lead federal agency for conducting R&D aimed at making solar PV more efficient and more affordable, NREL established itself as a leader in innovations across the entire spectrum of solar PV research, especially with second generation thin-film technologies and innovations to improve cell efficiency.

Yet NREL has played an equally important role in the indirect support of solar PV innovation. Crucially, it has acted as a public testing site of solar PV cell efficiencies, and the source of geographic assessments of potential levels of solar power production, thus providing essential enabling support across the entire solar PV innovation system. NREL is also the key federal partner in the Database of State Incentives for Renewables & Efficiency

(DSIRE), created in 1995 as "a comprehensive source of information on state, federal, local, and utility incentives and policies that support renewable energy and energy efficiency."[14] Given the many different sources of solar policy and incentives, DSIRE increases public acceptance of solar by acting as a potential buyer's clearinghouse for current information at one reliable source. Thus, it creates an even playing field for purchasers of solar and in turn, an environment in which market signals can be more easily recognized.

In 1996, solar PV activities at NREL were given greater focus and coherence through the creation of the National Center for Photovoltaics (NCPV), whose mission is to nurture and drive innovations that increase U.S. industry growth and competitiveness in solar PV manufacturing. NCPV pursues its mission by partnering with other organizations, facilitating technology transfer, serving as a clearinghouse for technology information and coordination, setting standards for measuring and testing new technology, and, most recently, supporting the activities of DOE's SunShot initiative.

The focus on public-private partnerships and market-based solutions remained central to government efforts throughout this era. Another initiative in this vein was DOE's 1997 Million Solar Roofs Initiative (MSR), to accelerate market acceptance of solar energy. During the eight years that the program was in place, nearly 1,000 private sector firms, utilities, developers, NGOs, and governmental agencies voluntarily signed onto the program with the goal of accelerating the adoption of solar installations. The purpose of the MSR initiative was to build a grass-roots network of technologically savvy people with valuable knowledge about local markets, rather than focusing purely on technology development. Additionally, this project produced best practices and a practical knowledge base that served as a basis for future market-based initiatives.[15]

Era IV: Stimulus, Reinvestment, and Recovery, 2006-2013

The shift of power in Congress following the 2006 elections led to a resurgence of support for renewable energy in government, and funding for solar PV research reached levels not seen since 1982, with R&D support rising from $65 million in 2006 to $148 million in 2007, and support staying at approximately this level ever since. This began the fourth era of solar PV innovation, marked by background forces of shifting political power followed by economic recession, stimulus, reinvestment, and recovery. Solar PV technology benefitted from these trends, being touted as part of a "green" economic stimulus.

This latest era of government support for solar PV innovation expands and extends the complex array of institutions and policy tools that evolved in the previous era. For example, starting in 2009 DOE sought to expand and deepen the capacity of local governments to adopt solar technologies through its partnership with 25 "Solar America Cities" around the country. As part of this effort, DOE compiled case studies and best practices to assist communities in implementing policies and standards for enhanced use of solar energy. Similar partnerships continue today at the state and local level all across the nation. [16]

Loan guarantees are another policy tool that DOE has increasingly used to support solar PV innovation. A Republican Congress gave DOE the authority to pursue such programs in the Energy Policy Act of 2005, with the intent of encouraging the federal government to support a suite of emerging technologies that were still too early in their market evolution to gain favorable financing deals from traditional sources of capital. A similar approach had already been in place for supporting the advance of nuclear power, and starting in 2009 DOE sought to provide loan guarantees that could reduce the risk for investors in renewable energy projects, including solar PV. Despite the well-publicized failure of a few of these loans (a not-surprising outcome given that the program is intended to support promising projects that are nonetheless too risky to be fully backed by private investors), overall performance has been strong, with only 6% of the $16 billion in loan guarantees associated with companies that filed for bankruptcy.[17]

Importantly, in this era DOE is also pursuing policy innovations in its more traditional support for fundamental science. For example, starting in 2009 it began funding new Energy Frontier Research Centers (EFRCs) to accelerate scientific discovery that would support new, disruptive technological and manufacturing breakthroughs. The EFRCs have been selected by scientific peer review and funded for five years at between $2 and $5 million per year. Major advances in third generation solar materials science, such as flexible, printable solar PV cells, have been produced at EFRCs and DOE-supported institutions.[18] The EFRCs include universities, national laboratories, firms, and non-profit organizations, often working in partnerships. There are currently 46 EFRCs in 35 states and the District of Columbia, with 115 participating institutions.[19]

Finally, the DOE's SunShot initiative was initiated in 2011 with the ambitious and comprehensive goal of making "...solar energy cost-competitive with other forms of electricity by the end of the decade." In contrast to the Million Solar Roofs project, the SunShot initiative includes a major push for technological innovation. This $145 million program is the largest single investment since the end of the first era that is focused primarily on solar technology (particularly solar PV). While it is far too early to judge whether SunShot will be able to meet its goals, the program in many ways draws upon what has been learned in a half-century of government experience with advancing solar energy, using a comprehensive and integrated array of approaches that range from support of fundamental science, to smart-grid research that will allow better handling of intermittent energy from renewable sources such as solar PV, to efforts to overcome policy barriers to integration of solar technologies into utility grids.

Status and Prospects for Solar PV

The benefits of solar PV to the U.S. economy continue to grow. In 2007, the estimated total solar workforce was almost 20,000 workers and that figure rose to nearly 120,000 in 2012. The five states with the most megawatts (MW) of installed solar capacity are California (2,902 MW), Arizona (1,097 MW), New Jersey (971 MW), Nevada (403 MW), and Colorado (270 MW).[20] Manufacturing of solar technologies has been strongest in states with strong historical industrial infrastructures such as Ohio. Recent plans for new manufacturing facilities in the western U.S., however, have generally fallen prey to the impact of China flooding the market with cheap solar modules.[21]

The domestic solar PV sector has seen spurts of growth since the first workable solar cells were invented in the 1950s. This is especially true in the past decade as U.S. installed solar PV increased from 138.8MW in 2000 to 7,700MW in 2012,[22] or about the capacity of 12 average-size coal-fired power plants. Prices for installed systems have also dropped dramatically from about $12/W in 1998 to just under $4/W in 2012. As prices decrease, demand for solar PV should remain strong. However, incentives are still a factor and installers express concern about their business's viability, given the unstable nature of government support.[23]

Lessons for Innovation Policy

Federal policy on R&D for solar PV has fluctuated along with changing political and economic forces and priorities. Government funding for solar R&D in the past half-century has been markedly inconsistent and never reached more than a fraction of the support consistently devoted to R&D for nuclear and fossil fuel technologies. At the same time, and partly in an effort to overcome this inconsistency, the government has adopted an increasingly diverse and comprehensive set of policies aimed at advancing solar PV.

From the earliest work on solar PV technologies, government has been a crucial contributor, beginning with R&D contracts that led to the formation of the solar business clusters in southern and northern California in the 1950s. Today, the improvements and increasing penetration of the technology are gradually allowing regional financial institutions and venture capitalists to become familiar with the solar industry, and thus more comfortable with managing the risk involved in such investments. But this is only now, slowly becoming a national trend. The continued support offered by programs such as the SunShot Initiative, loan guarantees, and tax incentives will likely be necessary to help reassure investors, local governments, small and large businesses, R&D institutions, and industry regulators alike, that solar is here to stay. Such reassurance will depend, above all, on the continuity of government commitments.

Government can act to mitigate uncertainties in the marketplace that impede innovation, but other forces will also be at play and cannot always be contained. From 2004-2010, the number of U.S. companies reporting shipments on solar PV cells and modules increased over 600%.[24] While the DOE's investments in spurring domestic market growth certainly influenced this trend, the price of silicon also had a big effect, being the single costliest component in silicon solar cells. As the demand went up during this time, prices briefly spiked. The biggest rise in shipments corresponded with a global shortage of silicon roughly between 2004 and 2007. In part, this shortage was driven by heavy foreign investments in solar installation, combined with continual demand for silicon-based electronics. For instance, Germany achieved a near ten-fold increase in their installed national solar capacity from 435 MW in 2003 to 4,170 MW in 2007. This exponential growth has continued and their installed capacity stood at 25,039 MW by the end of 2011.[25]

Silicon prices fell by more than 50% in just three years. China was found by the Commerce Department to have engaged in unfair business practices, flooding the global market with unusually cheap solar panels[26] that have driven down prices overall and this put several planned U.S. expansions or startups of domestic solar manufacturing (and hundreds of well-paying jobs) at risk.[27] To maintain private investments for innovation and domestic production, the federal government would have to support the stabilization of the solar PV supply chain, or at least protect the markets from unfair predatory practices of foreign actors. Such actions would require continued bold international trade diplomacy, in addition to continued robust support for the array of policies that have been pursued to date.

Conclusion

The ways in which government currently funds solar PV are much more diverse than in the 1950s and 1960s. Government investment in solar PV technology during the intervening decades can be credited for stimulating continual technological advance. The efficiency of solar cells improved as much as three-fold, in some cases capturing more electricity from the sun than was theoretically thought possible. Large-scale solar PV installations have made electricity cheaper by competing with expensive, variable cost peak generators. Small-scale rooftop solar PV installations have become efficient enough to supply nearly all the power needs for common suburban homes. Government funding has supported public and private R&D institutions and employed thousands of people in well-paying, domestic jobs. The installation of distributed rooftop solar continues to support local jobs in the construction sector as well. Government support for R&D has also contributed to the reduction in the market price of solar, making it competitive with traditional sources of electricity generation.

To maintain demand for solar PV products, and increase availability of affordable capital to solar startups, government has assumed risk through loan guarantees and grants. In the same way that government has used loan guarantees to back nuclear power generating facilities, it has also backed solar PV facilities. With sound supply chains, a robust market in which to sell products, and the opportunity to profit from their intellectual property, entrepreneurs have a greater incentive to innovate. The U.S. government has used diverse innovation policy tools to set the stage for increased innovation for solar PV technology as it competes with European and Asian countries, which are doing the same.

---

[1] Jessica Toothman and Scott Aldous, "How Solar Cells Work," HowStuffWorks website, available at: http://www.howstuffworks.com/environmental/energy/solar-cell1.htm

[2] Geoffrey Jones and Loubna Bouamane, "*Power from Sunshine*": *A Business History of Solar Energy* (Cambridge, MA: Harvard Business School, Working Paper 12-105, 25 May 2012).

[3] Phech Colatat, Georgeta Vidican, and Richard K. Lester, "Innovation Systems in the Solar Photovoltaic Industry: The Role of Public Research Institutions," presentation at the Stavanger Innovation Summit, Industrial Performance Center, MIT, 2009; Jones and Bouamane, *"Power from Sunshine."*

[4] "Best Research-Cell Efficiencies," National Renewable Energy Laboratory website, available at: http://www.nrel.gov/ncpv/images/efficiency_chart.jpg

[5] Lloyd O. Herwig, "Cherry Hill Revisited: Background Events and Photovoltaic Technology Status," *AIP Conference Proceedings* 462 (1999): pp. 785–795.

[6] All historical budget numbers are adjusted to reflect 2011 U.S. Dollars.

[7] Frank N. Laird, *Solar Energy, Technology Policy, and Institutional Values* (New York, NY: Cambridge University Press, 2001), p. 207.

[8] Ibid.

[9] B. Gallagher, P. Alexander, and D. Burger, *Electricity from Photovoltaic Solar Cells: Flat-Plate Solar Array Project Final Report, Volume V: Process Development* (Springfield, VA: JPL Publication, 86-31, NASA, 1986).

[10] Ibid.

[11] Ken Zweibel, "Thin Films: Past, Present, Future," National Renewable Energy Laboratory report, NREL/TP-413-7486 (1 Apr. 1995), available at: http://www.osti.gov/servlets/purl/61140-cYAVRN/webviewable/.

[12] Ibid.

[13] Jones and Bouamane, *"Power from Sunshine,"* p. 35.

[14] "About DSIRE Database of Energy Efficiency, Renewable Energy Solar Incentives, Rebates, Programs, Policy," DSIRE website, available at: http://www.dsireusa.org/about/

[15] G. Strahs and C. Tombari, *Million Solar Roofs: Final Report October 2006* (Washington, DC: U.S. DOE, Energy Efficiency and Renewable Energy, Oct. 2006).

[16] U.S. Department of Energy, *Solar Powering Your Community: A Guide for Local Governments*, 2nd Ed. (Washington, DC: U.S. DOE, Energy Efficiency and Renewable Energy, Jan. 2011).

[17] " Romney's Clean Energy Whoppers," FactCheck.org website (8 Oct. 2012), available at: http://www.factcheck.org/2012/10/romneys-clean-energy-whoppers/

[18] Jongseung Yoon, Sungjin Jo, Ik Su Chun, Inwha Jung, Hoon-Sik Kim, Matthew Meitl, Etienne Menard, Xiuling Li, James J. Coleman, Ungyu Paik, and John A. Rogers, "GaAs Photovoltaics and Optoelectronics Using Releasable Multilayer Epitaxial Assemblies," *Nature* 465, no. 7296 (20 May 2010): pp. 329–333.

[19] "Energy Frontier Research Centers (EFRCs)," U.S. Department of Energy website, Office of Science, available at: http://science.energy.gov/bes/efrc/

[20] "Solar Industry Data," Solar Energy Industries Association website, available at: http://www.seia.org/research-resources/solar-industry-data.

[21] Gary Nelson, "Delay of First Solar Plant Raises Fears Construction Continues, but Launch on Hold," *The Arizona Republic* (28 Apr. 2012).

[22] International Energy Agency, *Trends in Photovoltaic Applications: Survey report of selected IEA countries between 1992 and 2009* (Paris, France: International Energy Agency Report IEA-PVPS T1-19, 2010).

[23] Michael Sheffield, "Cutbacks to TVA's Solar Incentive Program Causing Concern for Some Installation Providers," *Memphis Business Journal* (27 July 2012).

[24] "Solar Photovoltaic Cell/Module Manufacturing Activities," U.S. Energy Information Administration website, available at: http://www.eia.gov/renewable/annual/solar_photo/archive/2009/

[25] Bundesministerium für Umwelt, Naturschutz, und Reaktorsicherheit, *Zeitreihen Zur Entwicklung der erneuerbaren*

Energien in Deutschland (Stuttgart, Germany: Zentrum für Sonnenenergie und Wasserstoff-Forschung, July 2013) available at: http://www.energiestiftung.ch/files/downloads/energiethemen-erneuerbareenergien/zeitreihen-zur-entwicklung-der-erneuerbaren-energien-in-deutschland.pdf

[26] Diane Cardwell, "A Tug of War Over Solar Tariffs," *New York Times*, Green blog (3 Oct. 2012), available at: http://green.blogs.nytimes.com/2012/10/03/a-tug-of-war-over-solar-tariffs/

[27] Ryan Randazzo and Gary Nelson, "First Solar to Lease Half of New Mesa Factory," *The Arizona Republic* (18 Oct. 2012).

7

ELECTRIC VEHICLE DEVELOPMENT AND BATTERY TECHNOLOGY

Eric B. Kennedy and Heather M. Ross

Introduction to Hybrid & Electric Vehicles

Electric vehicles are experiencing a resurgence in popularity among American consumers. This shift is the product of complex processes, including industry decisions, consumer choices, social influences, and the actions of government. A confluence of factors—including rising oil prices and geopolitical instabilities, a focus on developing a sustainable energy economy, new technological opportunities, and a desire to stimulate innovation and jobs at home—have propelled the government to support the electric vehicle (EV) industry in many different ways. Indeed, an examination of federal policy, stakeholder publications, and popular media coverage reveals an industry that has and will continue to be powerfully shaped by the involvement of government actors, just as the auto industry as a whole has been shaped since its inception.

The history of electric vehicles in the United States can be roughly divided into three phases: the first generation EVs at the turn of the 20th century; early consumer EVs (e.g., the GM EV1) and hybrids (e.g., the Honda Insight and Toyota Prius) introduced in the late 1990s; and the so-called "new-generation" hybrid and electric vehicles like the Chevrolet Volt and Tesla Roadster a decade later. A clear technological lineage exists between the second and third phases, although only gasoline hybrids were broadly available to consumers between the termination of the GM EV1 project in 2001 and the release of the first Tesla Roadster (2008) and Chevrolet Volt (2010). A complete timeline of all major electric and hybrid vehicles available to the public (as well as concept models) is maintained by the Union of Concerned Scientists.[1]

Hybrid and electric vehicles have received a great deal of focus and investment via federal and state policy. For the government, these partnerships represent more than simply investments in green technology; instead, they are interventions justified for a host of reasons that appeal to voters, from supporting cutting-edge research to protecting local jobs. The specific programs that impact electric vehicles are similarly diverse, ranging from consumer tax credits that directly support the purchasing of EVs, to improvements gained from investments in enabling technologies (primarily batteries but also lightweight materials and other components). Overall, the federal government has been involved in the development of electric vehicles in four key ways: funding research and development projects, directly and indirectly supporting the manufacturing of the vehicles, providing consumer incentives for adoption, and setting national emissions standards.

Hybrid & Electric Technology

Electric vehicles offer a technology that replaces vehicle emissions (carbon and other pollutants) concentrated on urban streets with reduced and relatively cleaner emissions from electric generating sites generally located far from urban centers (including zero-carbon hydroelectric, nuclear, and renewable sources).[2] The amount of emissions reductions depends primarily on the sources of electricity that are being used to charge the vehicle, although it is also easier to manage and clean emissions from "upstream" fossil-fuel power sources that are stationary and centralized than from innumerable individual vehicles.[3]

There are three major classes of contemporary high fuel-efficiency vehicles that are often referred to as "electric vehicles" or "EVs" by government and industry: plug-in electric vehicles, plug-in hybrids, and gasoline-powered hybrids that employ electric elements to improve their efficiency. Yet there are actually many forms of fully electric vehicles, including solar cars (which have experienced significant development via public-private-university partnerships like the university-based World Solar Challenge) and electric vehicles that are powered via induction from electrified roads (similar to electric streetcars, but instead being wirelessly charged from below).[4] These are strong examples of collaborative development of new technologies through research partnerships among federal, private, and educational organizations, and the technologies developed for such unconventional EVs often end up incorporated into improving the efficiency and range of more conventional models. In particular, these projects frequently couple relatively small investments (e.g., $2.7 million from the U.S. Federal Transit Administration to Utah State University to develop and implement an inductively charged bus prototype on campus) with industry and academic expertise to implement new ideas in practical demonstration projects.[5]

For plug-in electric vehicles (PEVs), power to the drivetrain is generated by an electric motor, which is in turn powered by onboard batteries. The vehicle itself can generate small amounts of power, primarily through regenerative braking technology. In this arrangement the momentum of the vehicle is slowed by the drag of the electric motors working instead as electricity generators, a method most effective in stop-and-go city driving. Much of the battery's energy, however, comes from charging stations located at the driver's home or work, or various public spaces. As the vehicle itself does not produce any direct emissions, many air quality policies consider it to be a zero emissions vehicle.

Plug-in electric vehicles, however, represent only a small part of EV development and commercialization. One of the first major contemporary forays into low emissions vehicles was through gasoline-electric hybrids, also known as hybrid-electric vehicles (HEV), like the Honda Insight and Toyota Prius. HEVs combine an electric motor and battery pack with a small, highly fuel-efficient gasoline-powered engine to run at its near-peak efficacy range whenever possible. These power sources may be combined in several ways. The car can be propelled by the electric motor alone, with a gas-powered generator providing supplemental electricity (known as a series hybrid). Alternatively, in a parallel hybrid, the combustion engine is used in conjunction with the electric motor to provide power to the drivetrain, with one motor providing the primary power while the other supplies extra power for climbing, high speeds, and acceleration. Some technologies, like the power split device found in the Toyota Prius, allow the engine and motor to act either in series or parallel depending on the usage.

An emerging version of the HEV is the plug-in electric hybrid (PHEV). A PHEV employs an onboard battery pack that is much larger and higher capacity than its HEV counterparts. This vehicle is plugged into a charging station when not in use. The larger reserve of battery power provides a longer range on electric power alone and greater flexibility in blending gasoline and electric power. In addition to using the engines in series or parallel like a HEV, PHEVs can be configured to travel for a limited range on electric power alone (e.g., 20 to 40 miles to allow for emissions-free, within-city use). Once the battery power is substantially reduced, the internal combustion engine can then provide additional range in series or parallel, achieving many of the benefits of a PEV but with a longer range than for electric power alone. These differences, however, make the PHEV more dependent on battery technology and potentially increase the cost of the vehicle because of the larger battery pack.

History of Electric Vehicle and Battery Technology Development

At the turn of the 20th century, electric cars offered lower noise and vibration, simpler maintenance, and easier starting systems than their internal combustion and steam counterparts. As a result, electric cars enjoyed a sizable market share. In 1912 nearly 34,000 electric vehicles were registered, accounting for 38% of the vehicle market in comparison to gasoline vehicles which accounted for only 20% of the market.[6] These early electric cars were widely preferred by women, as internal combustion engines required strenuous hand cranking. Ultimately, the development of the battery-powered automatic starter in 1913 led to the dominance of the internal combustion engine.[7]

Worldwide production of PEVs today, however, "remains at barely perceptible levels" compared to the nearly 80 million vehicles produced annually, and even HEVs account for less than 2% of global output.[8] In America, this translates to approximately 40,000 sales of PEVs since their introduction in 2010, or less than 0.3% of the 15 million light-duty vehicles sold during the same period.[9] According to the Congressional Budget Office, "most analysts expect sales of electric vehicles to grow modestly ... although such vehicles will probably continue to make up only a small share of total vehicle sales for many years to come".[10]

Contemporary development of fuel-efficient vehicles, including EVs, was largely triggered by concerns surrounding rising fuel costs and dependence on foreign oil during the 1970s oil shocks, despite the reality that the United States has some of the lowest gasoline and diesel prices among developed nations.[11] In 1976, in response to the energy crisis, Congress passed, over a presidential veto, the Electric and Hybrid Vehicle Research, Development and Demonstration Act of 1976 (P.L. 94-413), providing funding for research programs to develop lower cost, higher energy-density batteries for vehicles.[12]

The $160 million program funded automotive companies, battery manufacturers, and national laboratories, with a focus on improving lead-acid battery technology for powering vehicles beyond the automatic starter and auxiliary function power of batteries in internal combustion vehicles. The 1976 Act initially included an electric vehicle demonstration program, but this was canceled in 1981 in response to a GAO report that found the private sector demonstration premature.[13] However, the rest of the coordinated research and development program survived and was an important initial model for targeted energy research programs by enabling public-private partnerships that included national laboratories, automotive companies, battery companies, universities, and public and private energy companies.[14]

Yet it was not an American car company, but in fact the Japanese manufacturers Toyota and Honda that in 2000 made the first significant inroads into the America market with their high-efficiency HEVs. In this context, federal investment was important not only to developing the technology overall, but also in helping American automakers become more competitive in a U.S. market that had purchased 2 million HEVs by the end of 2011.[15]

U.S. efforts to advance EV and battery technology development and markets have been aided by four different avenues of government support. Through its various agencies, programs, and initiatives, the federal government has supported research and development projects for the component technologies required for PEVs, HEVs, and PHEVs. Additional support and investment has been directed at the manufacturing of these vehicles, including through direct financing, conditions on other funding programs, and moves to influence the priorities and expectations of automakers. National and state environmental programs have complemented these efforts through the implementation of emissions standards regimes that demand higher levels of efficiency from the auto market as a whole. Finally, governments have looked to influence consumer behavior primarily through incentive programs, but also through the use of other policy mechanisms.

Box 1. The Military and Early Efforts to Develop Battery Technology

Technological advances in transportation and military technologies in World War I altered the patterns of warfare, which in turn led to demands for increased portability and miniaturization of transceivers for battlefield communication during World War II. However, available battery technologies using a zinc-carbon platform relied on scarce metals for the manufacturing process with limited accessible supply due to wartime activities. Moreover, zinc-carbon designs were susceptible to cell destruction in high heat and humidity, meaning that a majority of batteries for troops in the Pacific and North African theatres arrived in a discharged and unusable state. The batteries that did work demonstrated a decline in discharge voltage nearing the end of life, and a resulting decline in function and range of communication equipment. [16]

In 1941, the U.S. Army Signal Corps called on inventor Samuel Ruben to improve small alkaline sealed cells with higher electrical capacity per unit volume, longer storage life in a variety of ambient temperatures, and constant discharge voltage.[17] Ruben's work led to a change from typical zinc-carbon cells to new mercury cell technology in 1942. The mercury battery became the standard for primary cells for the portable electronics industry. Ruben received royalties from the federal government for his work, but licensed manufacturing and marketing rights to the RM Mallory Company, with whom he maintained a collaborative relationship for more than 50 years.[18]

This relationship evolved in the context of the National Inventors Council (NIC), established in 1940 as a program of the Department of Commerce's Office of Technical Services. The chair of the NIC was Charles Kettering, director of research and development for General Motors. The NIC was a clearinghouse for American inventors whose work might have military application, in order to make connections between relevant inventions and the U.S. military. The work that Samuel Ruben and Mallory Battery Company did for the U.S. Army was the first significant public-private partnership in battery technology development. With early business activity with large battery orders placed during the war,

Ruben and the Mallory Battery Company (now Duracell), along with associated firms who were enlisted to assist with supply of the mercury cells during the war, capitalized on scientific advances realized as a result of the U.S. Army collaboration to continue to improve battery technology and significantly reduce costs for consumer applications. These developments helped enable the emergence of the portable transistor radio in the mid-1950s, and ultimately the feasibility of portable, miniaturized electronics more generally.[19]

Research & Development

Beginning around the 1990s, EV and battery research and development efforts ramped up in response to state and federal initiatives. Responding to environmental concerns, the State of California in 1990 passed a law that required a percentage of vehicles for sale in California to be "zero emission vehicles."[20] Shortly thereafter, Congress passed the Energy Policy Act of 1992 (P.L. 102-486), which included research, development, and deployment programs for fuel cells and electric vehicles, including $50 million over 10 years for demonstration programs, and an additional $40 million for infrastructure development for fuel cell and electric vehicle programs.[21] One notable component of the legislation mandated that at least 50% of funds for demonstration projects must come from non-federal sources, thus leveraging meaningful public-private partnership in electric vehicle research and development programs. In addition, the legislation placed a $4 million cap on any single project related to infrastructure development, to encourage widespread collaborative efforts with multiple partners.[22]

In 1991, Chrysler, Ford, and General Motors meanwhile formed the United States Advanced Battery Consortium (USABC) in partnership with Department of Energy, national laboratories, and several universities.[23] The USABC was targeted at developing electric and hybrid vehicle batteries. In 1992, the United States Council for Automotive Research (USCAR) was formed among the three major U.S. automakers.[24] Like USABC, the cooperation among firms and national laboratories that took place under USCAR was made possible by the 1984 Cooperative Research Act (P.L. 98-462), which allowed private sector firms to collaborate on research and development work that had previously been impossible due to antitrust regulations, a policy innovation that responded to mounting concerns about declining U.S. economic competitiveness in the 1980s.[25]

In 1993, the Clinton Administration moved beyond the public-private coordination model by expanding USCAR through the creation of the Partnership for a New Generation of Vehicles (PNGV).[26] PNGV comprised five federal agencies, along with USCAR.[27] The Department of Commerce was the lead agency in partnership with DOE, the Department of Transportation, the Environmental Protection Agency, and the National Science Foundation. Although federal money was not directly appropriated to PNGV, research agencies invested approximately $250 million per year, mostly via the DOE's energy conservation budget.[28] Auto industry partners contributed approximately $800 million per year. PNGV changed the public-private interaction from a focus on particular technologies to a more strategic partnership to improve fuel efficiency of conventional and new-technology vehicles. PNGV had three primary goals: improve manufacturing, implement new technologies, and develop a mid-sized passenger vehicle with approximately three times the fuel economy of a baseline 1994 passenger car, or 80 mpg.

The GM, Ford, and Chrysler concept cars funded through PNGV never reached market, though technologies resulting from the program, like regenerative braking and materials improvements, have contributed to performance improvement in today's vehicles. Although the federal investment was matched by industry-funded research, it is unclear whether the program resulted in research progress in any new domains or at a quicker rate than had industry pursued the projects on their own.[29] Furthermore, frustration emerged over the frequently changing approaches of federal sponsors, and resulting shifts in focus to pursue the latest research fads.[30] One significant finding from the PNGV experience was the realization that intellectual property requirements at corporate and federal levels were difficult to negotiate and contract, and were an obstacle to implementing research and development programs. As a result, USABC spent only $123 million of its $262 million budget from 1991-1995. This finding led to a recommendation from the GAO to revise federal requirements related to intellectual property in order to facilitate public-private innovation projects.[31]

Some of the research agenda of the PNGV program was shifted into the FreedomCAR project by the Bush Administration in 2002, then subsequently into the U.S. DRIVE program. This most recent iteration of these research goals (Obama Administration's U.S. DRIVE, or "Driving Research and Innovation for Vehicle efficiency and Energy sustainability") was announced in 2011, and is managed by the USCAR. Rather than providing funding, however, the DRIVE program consists of a set of technical teams, co-chaired by a DOE and industry representative, who meet monthly to "identify technology gaps, develop R&D targets, develop roadmaps to achieve technical targets and goals, and evaluate technical status and R&D progress."[32]

Currently, DOE has several offices involved in EV research, including the Applied Research Projects Agency - Energy ("transformational research on revolutionary, game-changing" technology), the Office of Science (basic materials research), and the Office of Energy Efficiency and Renewable Energy (applied research and development).[33] Contributions to the research and development of EVs also comes from grants made by other agencies (e.g., the National Science Foundation to universities) on complementary areas of basic and applied science, for example materials science that benefits vehicle components, aerodynamics studies that improve efficacy, and even research on batteries and motors from areas as apparently unrelated as space exploration.

Battery technologies have been an integral part of the DOE's research and development portfolio since the 1970s, and have featured prominently in many recent federal funding initiatives. In June of 2008, for instance, the DOE announced a $30 million investment to support PHEV projects pursued individually by General Motors, Ford Motor Company, and General Electric, with each company focusing on the goal of creating cost-competitive, 40-mile range battery vehicles for commercialization by 2016.[34]

In addition to partnerships with specific companies, DOE has established three major program foci for advanced battery technology for automobiles and transportation applications. Each focus area is centered at a designated national lab: advanced battery development, system analysis, and testing (USABC and the National Energy Technology Lab), applied battery research (Argonne National Lab), and focused fundamental materials research, or Batteries for Advanced Transportation Technologies (Lawrence Berkeley National Laboratory).[35] In addition to intra-agency collaboration, the Department of Energy coordinates with multiple national inter-agency and international collaborators including: Chemical Working Group of Interagency Advanced Power Group; Department of Transportation/National Highway Traffic Safety Administration; Environmental Protection Agency; United Nations Working Group on Battery Shipment Requirements; International Energy Agency's Implementing Agreement on Hybrid Electric Vehicles; 8-nation Electric Vehicle Initiative; and Clean Energy Research Center (U.S.-China).[36]

Through the Argonne National Laboratory, the federal government has sponsored over 45 national competitions to develop advanced vehicle technologies.[37] The most recent is the EcoCAR2 program, which represents a unique partnership between industry, the DOE, Natural Resources Canada, and student groups to pilot advanced PHEV technology. EcoCAR and EcoCAR2 were three-year programs (2008-2010 and 2011-2013) in which fifteen student teams across North America implemented PHEV modifications of conventional vehicles, in addition to developing outreach and business plans to build consumer interest in these technologies.[38] This kind of government investment not only results in the application of recent academic research in concept vehicles, but signals priorities to educational programs and generates skills and interest in electric vehicles among students in engineering and beyond.

In November 2012, the Department of Energy announced funding for the Joint Center for Energy Storage Research (JCESR), to be coordinated by and headquartered at Argonne National Laboratory.[39] This $120 million project was inspired by the collaborative lab environments of the Manhattan Project. It aims to increase the rate of development of battery technology and transfer from laboratories to the private sector for manufacturing and widespread adoption.[40] JCESR, which includes six national labs, five universities, and four private firms, expands the capacity for battery technology collaboration beyond application-specific technologies typically pursued by private sector firms. The goal of JCESR is to reduce battery costs five-fold and increase battery efficiency five-fold by 2018.[41]

The precise total dollar value invested in these numerous projects is difficult to determine because of a lack of consistent reporting across agencies, a tendency to report funding twice (once as part of an overall initiative, then again in individual programs or projects), and the diversity of agencies and programs contributing funding that ranges from direct investment in EV development (e.g., deployment projects) to indirect support of technologies that can contribute to EV innovations (e.g., basic material science). Yet, even focusing only on directly related efforts yields a large ongoing investment. According to the Breakthrough Institute, for instance, the overall federal funding for EVs and related battery technology between 2009 and 2014 is about $11.4 billion dollars.[42]

Supporting Manufacturing

Federal sponsorship of manufacturing and commercialization activities provides perhaps the most direct avenue of government investment in PHEVs and EVs. The American Recovery and Reinvestment Act (ARRA) of February 2009 made particularly large commitments, including $2.4 billion dollars dedicated to three of "the word's first electric vehicle factories" (operated by Nissan, Tesla, and Frisker Automotive[43]), in addition to another $2 billion in grants spread between 30 unnamed factories producing EV components including motors, drivetrains, and batteries.[44] For example, Navistar, a firm that received $39 million from ARRA grants, has since developed and marketed the only plug-in hybrid school bus, designed the first strictly electric heavy truck available commercially, and developed the first medium-duty commercial vehicle to be designated as a zero-emissions vehicle in California and a "clean fuel fleet vehicle" by the EPA.[45] The ARRA funding, through grants and loans, represents an important partnership with industry, as the companies have been required to provide matching investments, which have been applied to the development of manufacturing and EV battery recycling facilities.[46] Additionally, the federally funded Clean Energy Manufacturing Institute for Composite Materials and Structures may provide avenues to improve EV performance by reducing vehicle weight without increasing costs.

National Standards

Fuel economy standards have been another important policy tool for influencing innovation. First enacted by Congress in 1975, the Corporate Average Fuel Economy (CAFE) standards seek to increase the energy efficiency of cars and light trucks. While fuel economy rose significantly during the 1980s, Congressional action blocked increases in CAFE standards between 1990 and 2010, while allowing various classifications and exemptions (e.g., SUVs and minivans falling under less stringent "light truck" rules). As a result, fleet fuel efficiency of new vehicles was actually lower in model year 2007 than in 1987.[47] New standards set in 2012, however, require steadily increasing fuel efficiency through model year 2025 (40 mpg by 2021 and 50 mpg by 2025).[48] According to National Highway Traffic Safety Administration (NHTSA) estimates, while the fuel standards will add an average of $1,800 to the cost of new vehicles, they will end up producing a net lifetime saving of $3,400 to $5,000 due to reduced fuel expenditures. Because NHTSA only has the authority to set standards for five years at a time, official standards beyond 2017 will need to be adopted through future rulemaking.[49]

These standards can be directly linked to reductions—or at least slower overall national increases—in fuel consumption and to development of fuel-efficient technologies like EVs. According to data from the NHTSA and others, analyzed by the Union of Concerned Scientists, such standards have done a great deal to slow the rate of increase in total fuel consumption across America despite vastly increasing vehicle usage. While nationwide vehicle-miles-traveled increased by roughly 160% between 1970 and 2005, annual gasoline consumption increased by only 75% during the same span.[50] Furthermore, while hybrids may have played a relatively small part in achieving these nation-wide impacts because of their low adoption rate,[51] many of the efficiency technologies initially developed for EV applications have ended up being applied to improve the efficiency of conventional vehicles as well. Advances through EV research on drivetrain efficiency and reducing auxiliary loads, for instance, can be applied to conventional vehicles at relatively low cost to the manufacturer, resulting in fleet-wide efficiency.[52]

Because the CAFE standards set an average requirement for vehicle manufacturers, they do not translate into an equal increase in fuel efficiency across all of a company's vehicle models. Automobile manufacturers, for instance, may be able to continue to sell low fuel economy vehicles because of the success of their high economy vehicles, so long as the overall mix during a given model year conforms to the regulations.[53] Consider, for example, the EPA proposal to assign PEVs a rating of zero emissions per mile in some CAFE standards calculations because of a desire to increase their adoption. While this move might very well encourage the development of EVs (although many companies may have already decided to market EVs), overall it could mean that efficiency of companies' other vehicles can decrease substantially (e.g., because Nissan has already committed to produce up to 150,000 Leaf EV models, the remainder of their fleet's efficiency could worsen by up to 6 mpg while still meeting the CAFE requirements).[54] Alternatively, a company could aim to meet these standards by selling more efficient conventional vehicles, meaning they may have little impact on the production and sale of EVs specifically. Despite these tradeoffs, standards have and will likely continue to play a major role in signaling long-term regulatory trends to companies and consumers alike, thus motivating research, innovation, and production of EVs even if the short-term impacts are mixed and difficult to assess.

Action at the state level has played an important role in standards development as well. The California Air Resources Board in particular was a key driver behind the new CAFE standards by facilitating communication and collaboration between federal regulatory teams and industry executives. The combination of "a new generation of leaders taking over both the U.S. government and the global auto industry" and "dramatic shifts in public expectations for technology" occurred at the same time as states such as California moved to set their own standards for vehicle emissions and EV adoption.[55] Automakers' fears that state-by-state regulations would create an intractable patchwork of disparate requirements, as well as their weaker financial positions (and thus dependence on ongoing federal funding) resulted in relatively productive negotiations with the federal government leading to the latest standards.[56] The conditions for public-private cooperation in developing the new standards were thus clearly propitious. Nevertheless this case shows how government and industry can work together to create a broadly accepted yet ambitious regulatory framework.

Influencing Consumers

Among this wide variety of policy tools supporting EV development, direct incentive programs are perhaps the strongest mechanism for attempting to shape consumer behavior. The Clean Energy Act of 2005 provided tax credits of up to $3,400 per HEV purchased, for up to 60,000 cars per manufacturer, through 2010.[57] Beginning in 2008 with the Energy Improvement and Extension Act, and later in 2009 with the American Clean Energy and Security Act and ARRA, PHEVs benefited from similar tax credits, up to $7,500 depending on the vehicle and battery capacity. As of yet, the impact of these incentives on consumer behavior is difficult to judge. The Congressional Budget Office, for instance, suggested that tax credit incentives would need to be more than $12,000 to give consumers the same lifecycle costs as gasoline or HEV.[58] Yet consumers may have motivations beyond the purely economic for adopting PEV or PHEVs, whether because of environmental values or a desire to drive the latest technology. Research on incentive programs has generally found that gasoline prices may function as a stronger driver of consumer adoption of EVs than incentive programs, and that incentives providing up-front rebates provide more sway than later tax credits.[59] Additionally, these incentives can have a double effect: beyond just increasing the sales of PEVs and PHEVs, they can also trigger more competitive pricing of traditional, high-fuel-economy vehicles, thereby increasing sales of other low-emissions models as well.[60] According to the CBO, among policy alternatives, tax credits "are likely to have the greatest impact on vehicle sales" and on reducing gasoline consumption, continuing into the years after they expire because of their influence on the industry.[61]

The government also plays a role as a consumer, including through the federal vehicle fleet and military applications. ARRA, for instance, allotted $400 million for "transportation electrification demonstration and deployment" projects.[62] Although this money has funded numerous small-scale projects (many of which are local projects that provide free plug-in charging station installations to help encourage PEV adoption), it has had little impact on federal procurement.[63] Despite operating about 4,000 PEVs by the end of 2012, this represents less than 1% of the overall federal fleet, and seems to have been attained through an agency-by-agency piecemeal approach, rather than through a national acquisition program.[64] The Department of Defense has also had significant involvement in EV development both as a consumer and researcher. DOD procurement rules require "a preference for the lease or procurement of motor vehicles with electric or hybrid electric propulsion systems, including plug-in hybrid systems" (albeit excluding the majority of combat-related vehicles),[65] and base-specific projects (e.g. Los Angeles Air Force Base) have expedited the replacement of fleet vehicles with EVs.[66] Perhaps more interesting are projects that blend procurement with technology development, such as a new test project at Fort Carson that uses EVs to provide on-base grid stability.[67] Overall, such projects can advance DOD's core national security mission by enhancing grid independence and security on fixed facilities.

The federal government is not alone in working to build support for PEVs and PHEVs among consumers. Federal money has been used to support both state-level and nongovernmental partnerships to help support EV adoption. A $1 million investment, for instance, supported the Northeast Electric Vehicle Network, founded in 2011 and consisting of public representatives from several state governments.[68] The Network advocates for the adoption of cleaner vehicle technologies and the infrastructure needed to allow effective EV use. They also enlist support from private industry, such as an announcement in June 2012 of seventy companies (ranging from BMW to local industries) that support the adoption of EVs and the development of the related infrastructure through private fleet acquisition and advocating for charging station development.[69]

Current Status of Technology

HEV's accounted for about 1.6 million vehicles sold between 2005-2010.[70] However, despite significant investments in R&D, tax credits, and industry and infrastructure support, PEVs and PHEVs, which offer the greatest potential for switching away from fossil fuels, have not seen a substantial market share in the U.S. economy to date. Since 2010, only roughly 40,000 of the more than 15 million U.S. light vehicle sales have been PEVs or PHEVs.[71] This means that current tax credit schemes, which allow for up to 200,000 vehicles per manufacturer, have not come close to reaching their targets. Manufacturing capacity is strong, including DOE projections of about 500,000 batteries per year by 2015, which could drop manufacturing costs by an estimated 30-40%.[72] Indeed, if industry-provided numbers are to be trusted, capacity for the production of more than one million EVs by 2015 exists—the question is whether consumer demand will be able to meet this potential supply at current costs. General Motors, for instance, has reduced its projected sales for the Volt to instead "match supply with demand."[73] Various programs have been proposed to improve future prospects, including transforming tax credits into instant rebate programs.

Projects continue, however, to develop the infrastructure needed to sustain EVs at a national level. ARRA, for example, provided $400 million with a required match from private firms, to establish nearly 13,000 PEVs & PHEVs and over 22,000 charging points in over 20 focus cities around the country.[74] Such infrastructure is likely a prerequisite for significant expansion of the EV market.

Despite actual sales of individual vehicle models repeatedly falling short of projections, the overall markets for PEV and PHEV vehicles are growing and are expected to continue to do so.[75] This represents a sizeable market overall, with a value of $54 billion globally and $14.9 billion in the U.S. in 2011.[76] Foreign vehicle manufacturers continue to dominate the HEV market, with Chevy and Ford vehicles as only two of the top ten hybrids by sales in September 2012.[77] Yet, close to half of the top spots in the much smaller PEV and PHEV market are operated by American firms, primarily because of the presence of smaller firms like Tesla,[78] thus offering potential for U.S. advantage in the future.

Lessons Learned for Technology Policy

Federal involvement in the development of EV's has had complex motivations. Public and lobby pressure towards greening the transportation industry has certainly been an important factor, but it was a *convergence* of other factors—including the actions of individual states, existing research projects, and the economic downturn and resulting stimulus money, that led to the policy environment and technological prospects we see today. Similarly, the government's approach to funding EVs has been quite diverse, and has been framed as a way to address multiple national priorities (e.g., blending job creating and environmental investment), and to allow multiple federal agencies and programs to invest where they have expertise. There has been no overall national EV strategy; rather the federal government approach has responded to a *diverse set of priorities and opportunities* in making investments and setting policies.

There is some evidence that the benefits of federal involvement have been felt in these non-technical priorities as well, with employment in EV and battery-related fields increasing over 27% between 2007 and 2010.[79] The diversity of partners and approaches has supported both basic efficiency improvements and potentially radically innovative applications like using EV fleets as giant batteries to buffer power variation. However, federal involvement is also susceptible to competing priorities and industry influence (both positive and negative), as exemplified by the at-best modest success of the PNGV. Indeed, overall the degree to which direct federal investment in automotive companies has enabled EV projects that would otherwise not have been undertaken remains a subject of disagreement among analysts.

Many of the funding initiatives on EVs have been characterized by a *high degree of collaboration* between the government and private industry. While the automotive industry lobbied strongly against previous attempts to set fuel efficiency standards, a more collaborative approach has led to a more aggressive and long-term set of targets than many might have expected. The governments involved—be they federal, state, or municipal—have also used a *range of policy instruments* to catalyze the development of EVs. These tools span from directly funding research and development to setting mandatory regulations, and from formal, long-term planning to decisions as simple as allowing EVs to use carpool lanes - each of which can affect consumer and industry behavior. Furthermore, each individual policy instrument has a range of impacts (tax credits, for instance, may influence not only consumer purchasing behaviors, but also legislative perspectives on long-term priorities and directions[80]).

The federal government has played an enormous role in funding EV development, shaping research priorities, and guiding overall goals for the automotive industry. This is a policy arena where the federal government has been able to exert a large degree of influence over an emerging technology, and has played an essential role in its development. Moreover, in its efforts to advance EVs, the federal government has sought to advance multiple national priorities, ranging from building capacity in a nascent industrial sector, to improving environmental quality, national competitiveness, and energy security. While the specific impacts of particular investments and policies on the evolution of EV technologies may often be difficult to judge, on the whole it seems evident that the government is playing a positive and necessary role in the innovation journey of this complex technology.

---

[1] "Electric Vehicle Timeline: Electric Cars, Plug-In Hybrids, and Fuel Cell Vehicles," Union of Concerned Scientists website, available at:
http://www.ucsusa.org/clean_vehicles/smart-transportation-solutions/advanced-vehicle-technologies/electric-cars/electric-vehicle-timeline.html

[2] The exact improvements in net emissions depend largely on what fuel source provides the electricity, which varies substantially by region and electricity provider. See: Fong, Jocelyn, and Shauna Theel, "Myths and Facts about Electric Cars," Media Matters for America website (8 Feb. 2012), available at:
http://mediamatters.org/research/2012/02/08/myths-and-facts-about-electric-cars/185798

[3] Brent D. Yacobucci, *Advanced Vehicle Technologies: Energy, Environment, and Development Issues* (Washington, DC: Office of Congressional Information and Publishing, 6 Dec. 2007): p. 2.

[4] For public-private-educational partnerships to advance solar car design, see: the American Solar Challenge (http://americansolarchallenge.org/) or World Solar Challenge (http://www.worldsolarchallenge.org/). See note 5 for an example of inductive charging.

[5] This serves as an example of both an emerging technology and a government-educational partnership for conducting practical research. See: Jason Mick, "USU Researchers Trial Wireless Charging of Electric Bus," *DailyTech* (4 Dec. 2012).

[6] "Early electric automobiles," *Encyclopedia Britannica* website, available at: http://www.britannica.com/EBchecked/topic/44957/automobile/259061/Early-electric-automobiles

[7] Samuel Ruben, *The Evolution of Electric Batteries in Response to Industrial Needs* (Philadelphia, PA: Dorrance, 1978).

[8] Michael Renner, "Auto Production Roars to New Records," Worldwatch Institute, Vital Signs blog (11 Sept. 2012), available at: http://vitalsigns.worldwatch.org/vs-trend/auto-production-roars-new-records

[9] Congressional Budget Office, *Effects of Federal Tax Credits for the Purchase of Electric Vehicles* (Washington, DC: U.S. CBO special report prepared at the request of the Ranking Member of the Senate Committee on Energy and Natural Resources, Sept. 2012): p. 1.

[10] Ibid.

[11] Yacobucci, *Advanced Vehicle Technologies*, p. 2.

[12] Electric and Hybrid Vehicle Research, Development and Demonstration Act of 1976, Pub. L. No. 94-413, 15 U.S.C. § 2514 (1976).

[13] Bill Canis, "Battery Manufacturing for Hybrid and Electric Vehicles: Policy Issues," Congressional Research Service Report for Congress, Report 7-5700 (4 Apr. 2013).

[14] U.S. Department of Energy, *Electric and Hybrid Vehicles Program: 16th Annual Report to Congress for Fiscal Year 1992* (Washington, DC: U.S. DOE, Office of Transportation Technologies, Aug. 1993).

[15] "U.S. HEV Sales by Model," U.S. Department of Energy website, Alternative Fuels Data Center, available at: http://www.afdc.energy.gov/data/tab/vehicles/data_set/10301

[16] Eric S. Hintz, "Portable Power: Inventor Samuel Ruben and the Birth of Duracell," *Technology and Culture* 50, no. 1 (2009): pp. 24-57.

[17] Ruben, *Evolution of Electric Batteries.*

[18] Hintz, "Portable Power."

[19] Ibid. and Ruben, *Evolution of Electric Batteries.*

[20] California Environmental Protection Agency, "2003 Zero Emission Vehicle Program Changes," Air Resources Board Fact Sheet (18 Mar. 2004), available at: http://www.arb.ca.gov/msprog/zevprog/factsheets/2003zevchanges.pdf

[21] Ibid.

[22] Ibid.

[23] "Who We Are," United States Council for Automotive Research website, available at: http://www.uscar.org/guest/about/

[24] Ibid.

[25] National Cooperative Research Act of 1984, Pub. L. No. 98-462, 15 U.S.C. § 4301-4305.

[26] National Research Council, *Review of the Research Program of the Partnership for a New Generation of Vehicles: Seventh Report* (Washington, DC: National Academies Press, 2001).

[27] Brent Y. Yacobucci, "The Partnership for a New Generation of Vehicles: Status and Issues," Congressional Research Service Report for Congress, Report RS20852 (22 Jan. 2003).

[28] Ibid.

[29] Matthew Wald, "Hoping Not to Repeat the Mistakes of the Past," *New York Times* (31 Nov. 2008).

[30] Ibid. See particularly the quotation from Spencer Abraham.

[31] U.S. Government Accounting Office, *Electric Vehicles: Efforts to Complete Advanced Battery Development Will Require More*

Time and Funding (Washington DC, U.S. GAO/RCED-95-234, Aug. 1995).

[32] U.S. Council for Automotive Research, "U.S. Drive Partnership Plan," U.S. Department of Energy website (March 2012): p. 7, available at: http://www1.eere.energy.gov/vehiclesandfuels/pdfs/program/us_drive_partnership_plan_mar2013.pdf

[33] U.S. Department of Energy, "One Million Electric Vehicles by 2015," U.S. DOE Status Report (Feb. 2011): p. 7, available at: http://www1.eere.energy.gov/vehiclesandfuels/pdfs/1_million_electric_vehicles_rpt.pdf

[34] "DOE Announces $30 Million for Plug-in Hybrid Electric Vehicle Projects," U.S. Department of Energy website (12 June 2008), available at: http://energy.gov/articles/doe-announces-30-million-plug-hybrid-electric-vehicle-projects

[35] U.S. Department of Energy, *Energy Storage R&D: 2011 Annual Progress Report* (Washington, DC: U.S. DOE/EE-0675, 2012).

[36] Ibid.

[37] "Advanced Vehicle Technology Competitions for Students," Argonne National Laboratory website, available at: http://www.transportation.anl.gov/competitions/index.html

[38] Connie Bezanson, "EcoCAR 2: Racing Towards Vehicle Efficiency," U.S. Department of Energy website (23 May 2012), available at: http://energy.gov/articles/ecocar-2-racing-towards-vehicle-efficiency

[39] "Team Led by Argonne National Lab Selected as DOE's Batteries and Energy Storage Hub," U.S. Department of Energy news release (30 Nov. 2012), available at: http://energy.gov/articles/team-led-argonne-national-lab-selected-doe-s-batteries-and-energy-storage-hub

[40] Ibid.

[41] Benson Kong, "New JCESR Consortium Aims to Make U.S. a Battery Powerhouse," *Motor Trend* (3 Dec. 2012).

[42] Jesse Jenkins, Mark Muro, Ted Nordaus, Michael Shellenberger, Letha Tawney, and Alex Trembath, *Beyond Boom & Bust: Putting Clean Tech on a Path to Subsidy Independence* (Oakland, CA: The Breakthrough Institute, Apr. 2012).

[43] Ibid., p. 32.

[44] U.S. Department of Energy, "One Million Electric Vehicles," p. 5.

[45] "Sustainability," Navistar website, available at: http://www.navistar.com/navistar/whoweare/sustainability/products

[46] Office of the White House Press Secretary, "President Obama Announces $2.4 Billion in Grants to Accelerate the Manufacturing and Deployment of the Next Generation of U.S. Batteries and Electric Vehicles," press release (5 Aug. 2009).

[47] Union of Concerned Scientists, "The Costs of Delay: Fuel Efficiency and Auto Pollution Standards," fact sheet (Mar. 2011), available at: http://www.ucsusa.org/sites/default/files/legacy/assets/documents/clean_vehicles/Costs-of-Fuel-Efficiency-Delay.pdf

[48] Congressional Budget Office, *Effects of Federal Tax Credits*, p. 14.

[49] United States Congress, "2017 and Later Model Year Light-Duty Vehicle Greenhouse Gas Emission and Corporate Average Fuel Economy Standards; Final Rule," *Federal Register* 77, no. 199, (15 Oct. 2012), 62624 & 62627.

[50] Jim Kliesch, *Setting the Standard: How Cost-Effective Technology can Increase Fuel Economy* (Cambridge, MA: Union of Concerned Scientists, Sept. 2008): p. 5.

[51] Garth Heutel and Erich Muehlegger, "Consumer Learning and Hybrid Vehicle Adoption," John F. Kennedy School of Government, Harvard University, HKS Faculty Research Working Paper Series, RWP10-013 (April 2010). Note that, as referenced in this paper, several factors (including gas prices, vehicle cost, incentive programs, experiences with hybrids, and perceptions of the technology) can impact adoption rates.

Research on their relative importance is inconclusive and in early stages, according to the authors.

[52] Kliesch, *Setting the Standard*, p. 1.

[53] Congressional Budget Office, *Effects of Federal Tax Credits*, p. 14.

[54] "Designing a strong national clean vehicles program," Union of Concerned Scientists fact sheet, (30 Oct. 2009), available at: http://www.ucsusa.org/sites/default/files/legacy/assets/documents/clean_vehicles/designing-a-strong-national-clean-vehicles-program.pdf

[55] Neela Banerjee, "The Road to New Fuel Economy Standards was not Smooth," *Los Angeles Times* (18 Nov. 2011).

[56] Ibid.

[57] "Tax Incentives Information Center," FuelEconomy.com website, U.S. Department of Energy, available at: http://www.fueleconomy.gov/feg/tax_hybrid.shtml

[58] Congressional Budget Office, *Effects of Federal Tax Credits*, p. 12.

[59] David Diamond, "The Impact of Government Incentives for Hybrid-Electric Vehicles: Evidence from U.S. States," *Energy Policy* 37, no. 3 (2009): pp. 972-983; Ambarish Chandra, Sumeet Gulati, and Milind Kandlikar, "Green Drivers or Free Riders? An Analysis of Tax Rebates for Hybrid Vehicles," *Journal of Environmental Economics and Management* 60, no. 2 (2010): pp. 78-93; and David Browne, Margaret O'Mahony, and Brian Caufield, "How Should Barriers to Alternative Fuels and Vehicles be Classified and Potential Policies to Promote Innovative Technologies be Evaluated?" *Journal of Cleaner Production* 35 (2012): pp. 140-151.

[60] Congressional Budget Office, *Effects of Federal Tax Credits*, p. 12.

[61] Ibid., pp. iii & iv.

[62] "President Obama Announces $2.4 Billion in Funding to Support Next Generation Electric Vehicles," U.S. Department

of Energy website (19 Mar. 2009), available at: http://www.energy.gov/articles/president-obama-announces-24-billion-funding-support-next-generation-electric-vehicles

63 Diana Roycroft, "Fact Sheet: Plug-in Electric Vehicle Deployment," Environmental and Energy Study Institute website (2 Oct. 2012), available at: http://www.eesi.org/fact-sheet-plug-electric-vehicle-deployment-02-oct-2012; this source offers a substantial analysis of this investment.

64 "Federal Feet Report," U.S. General Services Administration website, available at: http://www.gsa.gov/portal/category/102859

65 Preference for Motor Vehicles Using Electric or Hybrid Propulsion Systems, Title 10 - Armed Forces, U.S. Code § 2922g.

66 Camron Gorguinpour and Arun Seraphin, "Air Force Jumpstarts Electric Vehicle Program," Office of Science and Technology Policy (9 Sept. 2011), available at: http://www.whitehouse.gov/blog/2011/09/09/air-force-jumpstarts-electric-vehicle-program

67 Pee Danko, "Electric Vehicles Recruited for Military Base Grid Security," *TG Daily* (30 Nov. 2012).

68 "Electric Vehicle Supporters," Transportation & Climate Initiative website, available at: http://www.transportationandclimate.org/content/electric-vehicle-supporters

69 "Private-Sector Companies Pledge Support for the Deployment of Electric Vehicles in the Northeast and Mid-Atlantic States," Transportation & Climate Initiative website, available at: http://www.transportationandclimate.org/private-sector-companies-pledge-support-deployment-electric-vehicles-northeast-and-mid-atlantic

70 U.S. Department of Energy, "One Million Electric Vehicles," p. 3.

[71] Congressional Budget Office, *Effects of Federal Tax Credits*, p. 1.

[72] U.S. Department of Energy, "One Million Electric Vehicles," p. 7.

[73] David Shepardson, "GM: August to be Volt's best month with sales topping 2,500," *The Detroit News* (29 Aug. 2012).

[74] Ibid., p. 5.

[75] "Electric Vehicles and Power Sources: Technologies and Markets," BCC Research website (Aug. 2010), available at: http://www.bccresearch.com/report/electric-vehicles-power-sources-fcb019d.html

[76] "Global & United States Electric Vehicles Market Forecast & Opportunities, 2017," TechSci Research website (Aug. 2012) available at: http://www.techsciresearch.com/1993

[77] Jeff Cobb, "September 2012 Dashboard," HybridCars website (4 Oct. 2012), available at: http://www.hybridcars.com/news/september-2012-dashboard-53157.html

[78] Jay Cole, "November 2012 Plug-In Electric Vehicle Sales Report Card," InsideEVs website (3 Dec. 2012), available at: http://insideevs.com/november-2012-plug-in-electric-vehicle-sales-report-card

[79] Jenkins, et al., *Beyond Boom and Bust*, p. 12.

[80] Congressional Budget Office, *Effects of Federal Tax Credits*, p. v.

8

BIOFUELS AND CELLULOSIC ETHANOL

Jen Fuller

Biomass, Biofuels, and Cellulose

The vast agricultural and forest resources of the United States make it a key player in the development of biomass as an alternative form of transportation fuels, with the potential for domestic and global impact. According to the U.S. Department of Energy (DOE), "The United States has over one billion tons of sustainable biomass resources that can provide fuel for cars, trucks, and jets; make chemicals; and produce power to supply the grid."[1] The U.S. Forest Service estimates that 8.4 billion dry tons of excess woody material must be removed from our forests to ensure their health.[2] Energy independence, energy security, stable jobs, and healthier communities are all motivations for advancing biofuel research, development, and deployment. By tapping into its domestic resources the United States is able to help meet its energy demands locally, while also keeping and building jobs in rural and urban communities across a wide range of industries. Petroleum costs are determined by a complex web of producers and futures markets which are outside the control of any one nation. However, when the United States invests in alternative sources of transportation fuels it can partially hedge against these market uncertainties by providing a domestic supply. Governmental policies and programs that encourage development and use of renewable fuels contribute to American innovations in technologies and processes with the potential for global application during this period of increased demand for oil and the decrease of easily accessibly transportation fuel reserves.

Biomass is a general term used to describe a wide array of plant and animal resources which can be converted into energy for heating, electricity, or transportation.[3] *Biofuels* encompass a range of inputs, or *feedstocks,* which can be broken down into three major categories: sugars and starch—including conventional feedstocks like the starchy, edible portion of corn as well as cane and beet sugars; lipids such as those secreted by algae and bacteria; and cellulose-based fuels like wood and agricultural residues. Biofuels are often discussed together without distinct separation. This report focuses in on *cellulosic biofuels,* such as switchgrass or poplar trees. Derived from the woody parts of plants, cellulosic biofuels have emerged as an alternative to both petroleum-based fuels and first generation biofuels like corn starch and sugar. Principal advantages of cellulosic inputs are the variety and relatively low cost of feedstocks, less intensive farming practices, and a lower greenhouse gas (GHG) footprint. According to the Energy Independence and Security Act (EISA) of 2007, cellulose-derived biofuel must achieve a 60% reduction in greenhouse gas emissions (from conventional fuels) over the entire life-cycle of the end product if it is to be used for vehicle fuel in the U.S.[4]

Concerns over conflicting land use for food and energy crops have also spurred research in the use of non-food biofuel crops to augment agricultural and urban residues. The goal is to reduce pressure on arable cropland and decrease global food price fluctuations that occurred as a result of crops, especially corn, being diverted for energy. Another impetus for research in advanced biofuels is to reduce GHG emissions produced by vehicles. Here cellulosic biofuels also appear to have a clear advantage over corn ethanol when considering the entire lifecycle from fertilizer inputs to harvesting, processing, transport, and end use emissions. For example, in a lifecycle analysis of multiple types of fuels, a National Academies study lists corn ethanol as creating just under 0.40 tonnes of carbon dioxide (CO_2) per barrel—a marginal improvement, at best, over gasoline, which produces 0.42. In sharp contrast, cellulosic ethanol produces net-negative lifecycle emissions of approximately -0.10 tonnes/barrel.[5]

Cellulose, together with hemicellulose and lignin, make up the strong structure of plant cell walls. Cellulosic biomass possesses a more complex and lengthier chain of molecules than starch (such as corn starch) or sugar cane, and in turn requires additional processes to access fermentable sugars. The resulting products include types of bioalcohol—typically in the form of ethanol—but other alcohols like methanol have been used in the past. *Bioenergy* refers to energy created from biomass including cellulosic biomass, while *bioproducts* refer to co-products created in the refining process.

Over the past nine years, the U.S. government has ramped up efforts to innovate along the entire cellulosic biofuel supply chain from optimizing and securing reliable feedstocks to encouraging domestic vehicle manufacturers to support engines designed for higher ethanol blends. Several agencies, national laboratories, numerous universities and research centers, and firms collaborate to conduct basic research, develop and test novel technologies and processes, and to scale up these activities from the bench to commercialization.

Private-sector involvement in commercializing cellulosic biofuels has to date been limited in part from concerns about lack of infrastructure, high production costs, and competition among a number of alternative fuel solutions.[6] However, mechanisms such as the Renewable Fuel Standard (RFS and updated RFS2), administered by the Environmental Protection Agency (EPA), are providing a stable market for biofuels despite the fluctuation of market prices for feedstock commodities. These demand-pull incentives also spur investment from petroleum companies.

Technological Background

This report focuses on ethanol derived from cellulosic biomass. Figure 1 illustrates the diversity of feedstocks, conversion processes, and products involved in the creation of biofuels. Reading this chart right to left: Wood, agricultural and marine waste to gasification/fermentation to bioalcohols illustrates the cellulosic ethanol journey, but it is just one of the primary pathways available for turning biomass into biofuels.

The area receiving the most federal investment is conversion processes due to the high cost of breaking down cellulose into accessible sugars. Humans have been converting sugars and starches into alcohols for thousands of years through fermentation—the process of yeast digesting sugars and expelling both CO_2 and forms of alcohol, like ethanol. Cellulosic biomass, however, requires additional processing in order to break down these molecular structures to access glucose—one of the simplest forms of sugar, present in all plants. Cellulosic biomass itself consists of three major components: cellulose, hemicellulose, and lignin. In order to ferment this matter into ethanol new forms of enzymes or yeasts must be developed to break down the hemicellulose.[7] This resistance to molecular decomposition has impeded widespread application until recently. There are two well-tested conversion pathways for creating cellulosic bioethanol: biological and thermal, while a third, pyrolysis, is still being developed. The basic elements of these processes are shown Box 1.[8]

Figure 1. Creating Biofuels

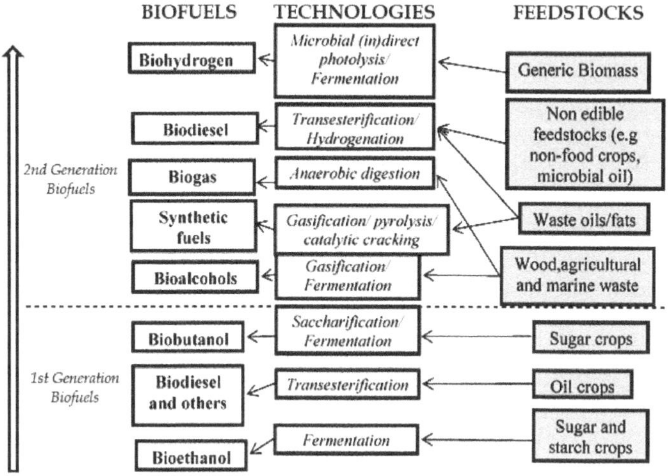

Source: Rafael Luque, Lorenzo Herrero-Davila, Juan M. Campelo, James H. Clark, Jose M. Hidalgo, Diego Luna, Jose M. Marinas, and Antonio A. Romero, "Biofuels: a technological perspective," *Energy and Environmental Science* 1, no. 5 (2008): pp. 513-596.

Box 1. Conversion Pathways for Creating Cellulosic Bioethanol

1. Biological Process

Pretreatment: The pretreatment process utilizes acids, ammonia, or other compounds to break down hemicellulose into "soluble five-carbon sugars—xylose and arabinose—and soluble six-carbon sugars—mannose, galactose, and glucose."[9] The specific sugars released depend on the available feedstock. Corn stover (leftover stems and leaves), for example, produces mainly xylose and glucose.

Conditioning: In this stage byproducts (like acid) from the pretreatment process which are harmful to the fermenting organism are removed and pH levels are restored to desirable levels.

Enzymatic Hydrolysis/Sacharrification: Enzymes such as cellulases or xylanases cleave cellulose into glucose and xylose. This process can take anywhere from hours to days depending on specific designs.

Biological Processing: Fermentation by modified yeast or proprietary enzymatic "cocktails" convert the broken down sugars into ethanol and its related co-products, carbon dioxide, and other chemicals depending on the feedstock and process pathways.

Chemical/Catalytic Processing: Chemical or catalytic cracking can be used after hydrolysis instead of, or in addition to fermentation for conversion into the desired fuel (ethanol, methanol, etc.).

Product Upgrade and Harvesting: This stage involves recovering co-products and separating fuel from water and any residual solids. Lignin is the bulk of the solids which can be used for heating and electricity (in integrated biorefineries), converted into intermediate chemicals, or changed into the precursors of other materials in the form of synthesis gas or pyrolysis oil.

2. *Thermochemical Process*

Feed Processing and Handling: For gasification to be effective feedstocks must be low in moisture and ash.

Gasification: Lignocellulosic feedstock is broken down thermally. This process involves using a gasifying agent like air, oxygen, or steam to produce raw syngas. A number of factors impact the quality and type of gas produced, such as feedstock, temperature, and pressure.

Gas Cleanup: Contaminants like tar, acid gas, ammonia are removed from the syngas while alkali metals are recovered.

Gas Conditioning: This step involves the reduction of hydrogen sulfide and adjustments of hydrogen-carbon monoxide ratios.

Fuel Synthesis: Synthesis gas composed of carbon monoxide and hydrogen is turned into alcohols or Fischer-Tropsch hydrocarbons. As with biological processes, heat can be captured as a co-product to increase the efficiency of operations.

Chemical/Catalytic Processing: Chemical or catalytic cracking can be used after hydrolysis instead of, or in addition to fermentation.

Balance of Plant: Considerations must be made for effective use of water, energy, heat and steam.

3. Pyrolysis

Pyrolysis is in the early stages of development and has the potential to meet longer term biofuels goals. A particularly promising component of this type of cellulosic biomass conversion is the wide range of intermediate products created including bio-oil, biochar (solid carbon rich product), and various forms of gases. These intermediary products can be transformed into bioalcohols (including ethanol), gasoline, diesel, jet fuel, chemicals, hydrogen, and can also be used for heating and electricity. Many of these compounds can be used in existing fossil fuel processing and distribution facilities.

Historical Roots: 1800 to the 1970s Oil Shocks

Biomass in the form of wood and other fodder has been used for millennia for heating purposes. Only recently has biomass been converted to liquid form for use in combustion engines. Because research in cellulosic biofuels has recently become a focus for the federal government, this section is divided into a historical perspective and a more detailed section on current activities.

In 1819 Henri Braconnet discovered how to break down cellulose into sugars using diluted sulfuric acid. In 1898, Germany utilized and later commercialized this process, producing 50 gallons per ton of biomass. Experimental fermentation stills in this early phase of development corroded easily because of the high amounts of acid used. In the early 1900s the United States adapted the German model to locally available feedstocks, resulting in two commercial scale plants, in Georgetown, SC and Fullerton, LA, using southern pine wood waste (i.e. shavings and sawdust) to create 5,000 to 7,000 gallons of ethanol per day.[10] DuPont purchased and operated the Georgetown plant until after World War I, and has remained interested in cellulosic fuel applications. Plants were also built to utilize paper black liquor (a byproduct of paper production) using yeast imported from Sweden, where the process was first developed. National research laboratories like the Forest Product Laboratory (of the U.S. Forest Service) contributed to these developments.[11]

Meanwhile, Rudolph Diesel, Henry Ford, and Nicholas Otto were building flexible fuel vehicles (FFVs) during the early days of automobile development in the late 1800s.[12] In 1908 Ford's Model T was designed as an FFV adjustable to alcohol, gasoline, or a "gasohol" mix. During the Second World War Germany, the United States, and Great Britain produced and used biofuels to assure adequate domestic fuel supplies. Research and development waned after the war owing to less urgent demand as well as falling gasoline prices.[13]

The history of cellulosic biofuel investment and support is linked to the history of ethanol and other biofuels more broadly. Between World War II and the next big push in development in the 1970s, ethanol was pursued for its anti-knock properties and as an alternative fuel by DuPont, General Motors Corporation, and other firms. The oil price shocks of the 1970s provided an inflection point in the development of biofuels. Brazil began to make use of their vast sugar plantations, and remains a key player in global ethanol research and deployment, while the U.S. relied on its well-established corn crops to create ethanol.[14]

1978 saw the first major ethanol legislation with the Energy Tax Act officially designating "gasohol" as a 10% blend and imposing a $0.04 gasoline tax, thus effectively providing a $0.40 subsidy per gallon of ethanol blended with gasoline.[15] In the 1980s, the United States began to phase out leaded fuel, increasing interest in alternative octane boosters (which allow higher compression of fuel without detonating) such as methanol and ethanol. Concerns over methanol-based additives (MTBE) due to ground water contamination also motivated a shift towards bio-based ethanol products. Up to this point, major research and development programs focused on corn starch and sugar based ethanol.[16]

The Government's Role: 2000 to the Present

More than twenty federal programs initiated over the past three decades have contributed to biofuels growth. These programs encompass all biofuel projects including ethanol, biodiesel, and jet fuel. This section looks at federal support for cellulosic ethanol in particular, but will also highlight broader reaching programs where appropriate.

The federal government has supported innovations in *cellulosic* biofuels along the entire supply chain, from inputs, processes, products, to developing markets and end users.

The Food, Conservation, and Energy Act of 2008 (known as the 2008 Farm Bill) approved $1 billion dollars in support for cellulosic biofuel development through 2012, primarily through the Department of Agriculture (USDA) and the DOE.[17] The Agriculture Act of 2014 (2014 Farm Bill) continues many of these programs, with $272 million in mandatory funding allocated primarily to cellulosic programs, with an additional $560 million in discretionary funding through fiscal year 2018.[18] DOE also provides ongoing support: since the 1970s, DOE has invested more than $3.63 billion for research, development, and demonstration in biofuels, with a particular emphasis on cellulosic biofuels, about $884 million of which is authorized under the American Recovery and Reinvestment Act of 2009.[19] Key programmatic aspects of these investments are described below, organized by stages of a simplified supply chain, including feedstocks, conversion, and products/end users.[20]

Most programs cut across these stages, but the supply chain format illustrates the complex range of activities and actors involved in cellulosic biofuel production including research in molecular biology, biochemistry, and forestry, and involvement of seed companies, farmers, transportation businesses, conventional oil refineries, chemical manufacturers, and vehicle manufacturers. Federal efforts have led not only to collaboration among agencies, but also among private businesses and universities in order to create a high level, integrated infrastructure for development of biofuels that would be difficult to achieve for the biofuel industry alone.

The Biomass Research and Development Initiative (BRDI) coordinates efforts across the eight federal agencies involved with biofuels research, and provides analysis on economic feasibility, sustainability, feedstock development, and conversion processes. The BRDI also maintains a useful glossary of common biofuel-related terminology.[21] The BRDI supports cellulosic biofuel research through R&D, grants, and loans. Over $300 million was allocated to the program from 2002-2012.[22] The 2014 Farm Bill continues mandatory funding at $3 million per year for FY 2014-2017, and discretionary funding continues at $20 million per year for FY 2014-2018.[23]

Feedstocks

The development of specific feedstocks as well as the harvest and transportation of these crops to biomass conversion facilities is supported through several of programs spread across the DOE, USDA, National Science Foundation (NSF) and other federal agencies. Research universities are partnering with the National Labs to find mechanisms for increasing ethanol conversion potential in organisms. Forests and Rangelands (a collaboration between the USDA and the Department of Interior) is seeking ways to collect excess biomass from public lands to be used in cellulosic biofuel production.[24] Farmers are key participants in these activities by field testing new crop varieties and growing, harvesting, and facilitating delivery of feedstocks to biorefineries.[25] Table 1 illustrates the diversity of feedstock options available for cellulosic biofuel conversion.

Cellulosic crops have several distinct advantages over corn or sugar feedstocks. The majority of the crops are perennial, meaning that their root structures remain intact throughout the year, aiding with the prevention of soil erosion and reducing the need for planting every season. These crops also require less fertilizer and less tillage of the land, which reduces fossil fuel use. However, the bulk and wetness of cellulosic feedstocks remains a challenge for efficient transportation.[26]

Table 1. Feedstock Options for Cellulosic Biofuels

Category	Type
Agricultural and Forest Residues	• Corn stover/sugar bagasse (e.g. stalks and leaves left over from harvest) • Cereal straws (e.g. wheat, oat, barley, and rice) • Sawdust and hulls • Fuel wood thinned to prevent fires
Urban Residues	• Construction and demolition site wood residues • Yard waste • Paper (black liquor) sludge
Energy Crops	• Trees (hybrid poplar, willow, silver maple, eucalyptus, and black locust) • Grasses (switchgrass, sorghum, reed canary grass, miscanthus)

Several challenges exist for advancing cellulosic feedstock availability. Crops require novel harvesting techniques in order to maximize both extraction for energy use and preservation for soil vitality. Percentages of crop retention vary according to crop species and the ecosystem in which that crop exists. Data is still being gathered on the best strategies for the harvest, transport, and storage of cellulosic feedstocks. Biorefineries are designed to operate continuously throughout the year; refineries in northern climates must therefore determine the most effective method for the harvest and storage of feedstocks during the winter months.[27]

Conversion Processes and Producers

Much of the federal support for advanced biofuels is related to developing new processes for converting cellulose, hemicellulose, and lignin into ethanol or other biofuels. Such new processes will require novel enzyme combinations for breaking apart the complex sugars embedded in cellulose and hemicellulose, new forms of gasification techniques, and pyrolysis development. In addition to specific conversion pathways, EPA, DOE, USDA, and other agencies also conduct numerous evaluative studies on the effectiveness of biorefineries and their related supply chains so that any improvements can be made as the refinery programs are scaled up to commercial applications.[28] This section discusses some of the federal programs that support basic research in enzyme combinations, as well as programs to advance biofuel production, including biorefineries.

The 2008 Farm Bill included incentives for producers of cellulosic ethanol including: The Biorefinery Assistance program (loan guarantees); Repowering Assistance (payments); Bioenergy for Advanced Biofuels (payments based on yearly increases in the production of cellulosic biofuels); and the Biomass Research and Development Initiative (R&D and demonstration of "biofuels and biobased products, and the methods, practices, and technologies for their production").[29] The Farm Bill also supports producers with a $1.01 tax incentive per gallon for ethanol created using cellulosic feedstocks. The provision was extended during the summer of 2012 and is expected to cost about $59 million over the next ten years.[30]

The primary challenge for conversion processes is to become more economically competitive with gasoline and corn ethanol. Basic research in molecular biology is needed to determine the underlying mechanism for resistance to biological or chemical decomposition.[31] Cellulase enzymes remain a significant portion of cellulosic ethanol production costs. As shown in the DuPont-NREL partnership described in Box 2, the federal government provides expertise and funds for basic research in enzyme optimization. Innovations in materials science will also prove beneficial for reducing conversion costs because pretreatment containers must be strong enough to withstand high temperatures and caustic matter.[32]

Box 2. Innovation Spotlight

Starting in 2000, DOE's National Renewable Energy Lab (NREL) and DuPont, along with other private sector participants and universities, have been collaborating to lower the cost of cellulosic ethanol production. DOE has provided $49 million in grants to develop pretreatment processes and improve the blend of conversion organisms used to break down the strong cellulosic material.[33]

Since 2003, DuPont has led a team under a DOE cost-shared cooperative agreement with NREL, partners from Michigan State University, and the firm Diversa, to improve the corn stover conversion process through increased efficiency and decreased costs.[34] Pretreatment innovations developed by DuPont used dilute ammonia in place of established acid processes to reduce cost and increase the efficiency of the glucose and xylose conversion rates. Diversa lent "enzyme cocktail" research experience to the project's saccharification objectives and continues to combine enzymes (like *Zymomonas mobilis*) in search of optimizing the process.

This partnership led to a successful pilot project, and further research was carried out in order to prepare these combined processes for commercial applications. DuPont and Genencor have invested $144 million to apply the enzyme and pretreatment processes developed in the NREL partnership to corn stover and sugar cane bagasse (stalks) based cellulosic ethanol.[35]

Basic knowledge on enzymes and conversion processes is now being used in the demonstration-scale cellulosic biofuel plant in Venore, Tennessee, which has been producing around 250,000 gallons of ethanol per year since December 2009.[36] The endeavor has also been supported by the state of Tennessee with a $73 million investment. This support helped the Oak Ridge National Lab secure an additional $140 million from DOE for the creation of a Bioenergy Science Center, one of three new centers nationwide.[37] The DuPont biorefinery in Venore, TN, which currently processes switchgrass, corn stover, biomass sorghum, sugar cane bagasse, and wood, was recently sold to TennEra, a company wholly owned by the University of Tennessee Research Foundation.[38]

The lessons learned from the demonstration plant will be applied to DuPont's latest commercial cellulosic ethanol endeavor, which will produce 30 million gallons of corn stover ethanol in Nevada, Iowa. Construction, feedstock storage, and permanent employee hiring are currently underway.[39]

Products and End Users

In addition to bioethanol there are numerous products and processes that can potentially be derived from cellulosic biomass conversion: cellulosic biodiesel and jet fuel, plastics, chemicals, and other co-products, fuels, electricity, and heat. Thus, a variety of private sector investment opportunities and market niches may be served by cellulosic biofuels. While these other products are an integral part of the biofuel conversion process, this section discusses the main product, cellulosic ethanol, in relationship to fuel standards, high ethanol blend vehicle incentives, and support for the supply of ethanol blends.

Fuel standards represent some of the longest-standing federal efforts to move the nation away from petroleum-based fuels. Most motorcycles, cars, and light trucks sold within the U.S. since the 1970s have the capability to run on ethanol/gasoline blends of 10%. There are also 62 2012 models of flex fuel vehicles (FFVs) available for purchase in the United States, and these are able to use up to 85% ethanol (E85) blends. By comparison, 31 hybrid models and 6 fully electric models were offered on the market in 2012.[40] Although the cost of E85 fuel is still relatively high (approximately $4.91 per gallon compared with $3.82 for gasoline in October, 2012[41]), engine compatibility is higher and more economical than other engine conversion processes, like compressed natural gas. According to DOE's Alternative Fuels and Advanced Vehicles Data Center, there are 2,264 U.S. gas stations that support an E85 blend,[42] about 1.5% of the 160,000 gasoline stations nationwide.[43]

Government is also playing a key role in boosting long-term demand for and private sector involvement with biofuels through vehicle procurement. The 1992 Energy Policy Act required governmental fleets of light vehicles and trucks to comprise a total of between 75% and 90% of E-85-compatible flexible fuel vehicles (FFVs). The requirement can also be met through implementing petroleum reduction programs (e.g., purchasing hybrid or electric cars or through actual usage of bio-based products).[44] Under the Act, 116,224 E85 flexible fuel vehicles were purchased from 1992 to 2010.[45]

The Renewable Fuel Standard (RFS) has been a major driver in the development of cellulosic biofuels. Parties required to participate in the RFS program, including refiners, importers, and blenders, are given a percentage of biofuels to be blended into their product—the Renewable Volume Obligation (RVO).[46] Shown in Figure 2, the percentage of cellulosic blends in the biofuels mix increases over time, and becomes the largest type of alternative fuel in the mix by 2022. The RFS and the RFS2 (updated in the Energy Independence and Security Act of 2007) also contain a requirement for biofuel life-cycle analysis from the field to tailpipe emissions. The emissions reduction schemes depend on the type of fuel but require cellulosic biofuels to achieve a 60% reduction from conventional gasoline. The RFS2 also requires land use and land cover analyses to determine the carbon dioxide impacts of the crop in relationship to land.

Figure 2. Renewable Fuel Standard Volumes by Year

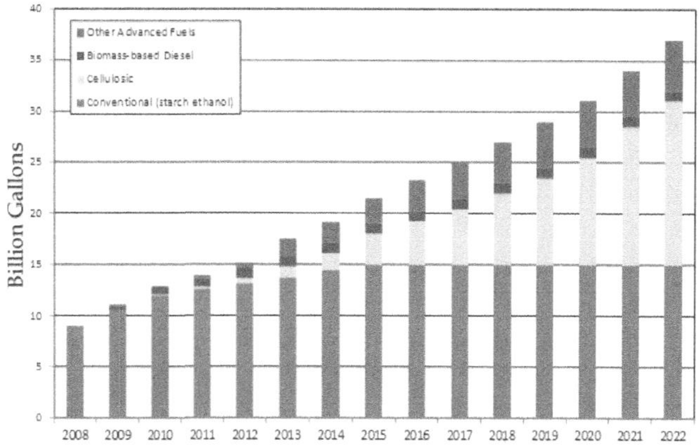

Source: U.S. Department of Energy, Alternative Fuels Data Center.

Combined with efforts to promote the use of FFVs and high ethanol blends, critical infrastructure measures are also imperative for a creating a new farm/forest-to-fuel system. Tax credits through the Energy Policy Act of 2005 are available to incentivize alternative fueling equipment for 30% of the cost up to $30,000 for businesses and $1,000 for individuals.[47] This provision is for commercially available alternative fuels including E85. While virtually all E85 is currently corn-starch based, as shown in Figure 2 above, the mix of cellulosic biofuels will rise and be compatible with the existing starch ethanol infrastructure.

In order to meet the RFS, infrastructure upgrades such as increasing number of vendors who carry higher ethanol blends (of up to E85) must continue to be supported. Markets for ethanol blends are tied to prices for feedstocks and oil. When gasoline prices are high, ethanol blends become more attractive. There are many flexible fuel vehicles available for purchase in the United States, so vehicle availability remains strong.

Cellulosic Biofuel Status, Markets, and Projections

Because large-scale deployment of cellulosic biofuels is still evolving, it remains to be seen how this product will impact markets or specific percentages of the transportation sector's energy usage. Cellulosic biofuel still accounts for only a small proportion of total U.S. biofuel production. In 2011, biofuels contributed about 14.8 billion gallons, or around 10.6%, of total transport fuel consumption. Corn starch ethanol accounted for 98% of this total, with much of the rest derived from vegetable oil, waste oil, and animal fat based biodiesel.[48] As of June 6, 2014, the United States has produced 515,563 gallons of cellulosic biofuel under the tracking process of the RFS2.[49]

If the RFS2 targets are fully implemented, crude oil imports will decrease as much as 17% by 2022,[50] and biofuels may take up much of the slack. The DOE/USDA Biomass Multiyear Program is an overarching plan to develop biofuels. Its goal is to apply R&D to making "cellulosic biofuels competitive with petroleum-based fuels."[51] The Energy Information Administration (EIA) predicts that the DuPont and POET/DSM biorefineries in Iowa, as well as Abengoa's plant in Kansas (which starts production in 2014), offer the best prospects for commercial-scale cellulosic biofuel production. The POET plant, built in conjunction with the Dutch company Royal DSM, opened in Emmetsburg, IA on September 3, 2014. The plant, named Project Liberty, was created with approximately $100 million from the DOE in cost-sharing support, along with nearly $20 million from the state of Iowa for construction and job support. Project Liberty is the second commercial-scale facility to open in the U.S. following the INEOS Bio plan in Florida, which opened in 2013.[52]

In this early stage of development, not every refinery, enzyme, or process will ultimately become commercialized. The federal government has invested in a variety of initiatives across the country in order to develop supply chains appropriate to surrounding resources and to ensure that there is a sufficient diversity of projects and approaches to assure overall success.

Feedstock Markets

Feedstock markets for cellulosic biofuels are difficult to chart owing to the diversity of crops and conversion processes. Evaluating cellulosic feedstocks compared to other biofuels is inevitably uncertain—for instance, forest resources can be cheaper at the outset, but may entail more costly transportation obstacles.[53] Despite such complexities, the abundance and diversity of feedstocks across the United States means that there should be a relatively stable supply of cellulosic feedstocks well into the future as the RFS increases.

Biorefineries

While the commercialization of cellulosic ethanol has fallen short of targets laid out in EISA, the federal government's strategic and integrated approach to advancing partnerships for biofuels development has made great strides towards creating a robust bio-based farm/forest-to-fuel system. Early studies suggest that biorefineries locate near feedstock sources to ease transportation logistics and to ensure a steady supply. Figure 3 shows the integrated biorefineries utilizing agricultural residues, woody biomass, and vegetative and yard waste. The

American Recovery and Reinvestment Act (ARRA) funded several integrated biorefinery projects, and many of these projects have secured additional funding from private venture capitalists and are continuing or have moved into collaborative research with public universities.[54]

Figure 3. Integrated Biorefineries in the U.S.

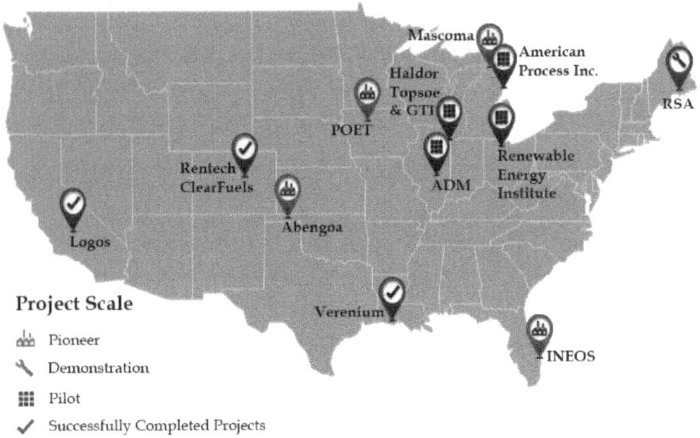

Source: U.S. Department of Energy, Office of Energy Efficiency and Renewable Energy.

Lessons Learned

Biofuels policy at the national level supports cellulosic ethanol production from laboratory testing (partnerships for creating enzymes and modification of crops) to the farm (conducting crop suitability testing; developing new harvesting techniques) to application in biorefineries (financial incentives and loan guarantees for construction and production) to the end user (flexible fuel vehicle requirements and cellulosic biofuel mandates). This integrated approach combines the expertise of multiple stakeholders to create a set of best practices for moving the technology forward. A benefit and a challenge of cellulosic biofuel lessons deals with the diversity of feedstocks. Different crops require different enzymes to break down the cellulose—meaning biofuel practices are not "one size fits all" solutions, but that numerous opportunities for innovation can be explored simultaneously.

The broadest reaching and most enduring programs seem to be those with wide collaborative networks across private industry, universities, and the federal government. Two examples of these collaborations are the Boardman ZeaChem facility, with a consortium based at the University of Washington; and the DuPont-University of Tennessee-NREL partnership's biorefinery in Venore, TN, which has developed considerably since NREL and DuPont began developing new conversion processes 10 years ago. Both of these projects developed wide networks in optimizing feedstock production and logistics, and enzyme and other conversion mechanisms, and both have resulted in a biorefinery. Specific federal mechanisms which supported these initiatives include ARRA grants, loan guarantees, partnerships with NREL and Oak Ridge National Laboratory, and other national partnerships. Washington State University, for instance, is researching poplar trees to increase their productivity for the ZeaChem biorefinery, and is a part of the National Advanced Biofuel Consortium. Crucially, these projects draw on support from not only the United States government, but also from their home states, private businesses, and universities.

Developing critical alternative fuel production capability and infrastructures from feedstock to fuel requires the cooperation of multiple institutions and sectors, and support from the federal government. This support is manifested in a myriad of ways and includes loan guarantees, grants, cost-share partnerships, tax incentives, research and development, high level programmatic assessment, and life cycle analyses. Today's energy problems require a multi-faceted approach in developing a number of alternative solutions. No single solution or entity will shift our energy use from primarily imported fossil fuels to renewable, alternative fuels such as cellulosic ethanol. Strong partnerships throughout the nation in a variety of fields can ultimately develop a cellulose-based biofuel industry as part of the solution to a host of energy challenges.

---

[1] The use of this term stems from the Billion Ton Update study. Sustainability within that report involves maintaining soil quality and using renewable, non-food inputs, which emit significantly less greenhouse gasses than conventional transportation fuels over the life cycle of the product; U.S. Department of Energy, *Biomass Multiyear Program Plan* (Washington, DC: U.S. DOE, Energy Efficiency & Renewable Energy, Apr. 2012).

[2] "Forest Biorefinery - Introduction," Forest Products Laboratory website, U.S. Department of Agriculture, available at: http://www.fpl.fs.fed.us/research/research_emphasis_areas/introduction.php?rea_id=3

[3] For more information on the complete range of biomass inputs and purposes, see: U.S. Department of Energy, *Biomass Multiyear Program Plan.*

[4] Energy Independence and Security Act of 2007. H.R. 6. 110th Congress., 1st sess. (2007).

[5] National Research Council, *America's Energy Future Panel on Alternative Liquid Transportation Fuels* (Washington, DC: National Academy of Sciences, 2009): p. 250.

[6] U.S. Department of Energy, *Biomass Multiyear Program Plan,* pp. 1-5.

[7] Robert F. Service, "Biofuel Researchers Prepare to Reap a New Harvest," *Science* 315 (2007): pp. 1488-1491.

[8] For detailed explanations of the various transformational process innovations, see: Rafael Luque, Lorenzo Herrero-Davila, Juan M. Campelo, James H. Clark, Jose M. Hidalgo, Diego Luna, Jose M. Marinas, and Antonio A. Romero, "Biofuels: a technolog-

ical perspective," *Energy and Environmental Science* 1, no. 5 (2008): pp. 513-596.

[9] U.S. Department of Energy, *Biomass Multiyear Program Plan*, pp. 2-30.

[10] E.C. Sherrard and F.W. Kressman, "Review of Processes in the United States Prior to World War II," *Industrial and Engineering Chemistry* 37, no. 1 (1945): pp. 5-8.

[11] Ibid.

[12] Luque, et al., "Biofuels," pp. 513-596; Barry D. Solomon, Justin R. Barnes, and Kathleen E. Halvorsen, "Grain and Cellulosic Ethanol: History, Economics, and Energy Policy," *Biomass and Bioenergy* 31 (2007): pp. 416–425; Service, "Biofuel Researchers," pp. 1488-1491.

[13] Solomon, et al., "Grain and Cellulosic Ethanol," pp. 416-425.

[14] Ibid.

[15] Ibid.

[16] Samuel B. McLaughlil and Lynn Adams Kszos, "Development of Switchgrass (*Panicum virgatum*) as a Bioenergy Feedstock in the United States," *Biomass and Bioenergy* 28 (2005): pp. 515-535.

[17] Kelsi Bracmort, Randy Schnepf, Megan Stubbs, and Brent D. Yacobucci, *Cellulosic Biofuels: Analysis of Policy Issues for Congress* (Washington, DC: Congressional Research Service, RL34738, 14 Oct. 2010), p. 19.

[18] Randy Schnepf, *Energy Provisions in the 2014 Farm Bill (P.L. 113-79)* (Washington, DC: Congressional Research Service, R43416, 12 Mar. 2014).

[19] Note that these funding levels may have overlap. All dollar expenditure amounts reported in 2011 constant dollars; U.S. Department of Energy, *Report on the First Quadrennial Technology Review: Technology Assessments* (Washington, DC: U.S. DOE, Aug. 2012), available at:

http://energy.gov/downloads/quadrennial-technology-review-august-2012

[20] More detailed analyses of the agricultural practices, harvesting, and transportation of feedstocks can be found in U.S. DOE, *Biomass Multiyear Program Plan.*

[21] The BRDI is composed of The Departments of Agriculture, Energy, Defense, Transportation, and Interior; The Environmental Protection Agency; The National Science Foundation; and the Office of Science and Technology Policy; Biomass Research and Development Initiative.

[22] Bracmort, et al., "Cellulosic Biofuels."

[23] Schnepf, *Energy Provisions.*

[24] "About Forests and Rangelands," Forest and Rangelands website, available at:
http://www.forestsandrangelands.gov/about.shtml

[25] U.S. Department of Energy, "Feedstock Supply and Logistics: Biomass as a Commodity," Energy Efficiency and Renewable Energy, Bioenergy Technologies Office summary paper (May 2013), available at: http://www1.eere.energy.gov/biomass/pdfs/feedstocks_four_pager.pdf

[26] G. Philip Robertson, Douglas A. Landis, and Madhu Khanna, "The Sustainability of Cellulosic Biofuels," Ecological Society of America summary paper (2012), available at: lter.kbs.msu.edu/wp-content/uploads/2012/06/sustainability-of-cellulosic-biofuels.pdf; Elias J. Halvorson, *A Century Long Pursuit of Alternative Fuels and Feedstocks: A Content Analysis* (Wright-Patterson AFB, OH: Air Force Institute of Technology, Master's Thesis, 2011).

[27] U.S. Department of Energy, *Biomass Multiyear Program Plan.*

[28] Ibid.

[29] Bracmort, et al., "Cellulosic Biofuels."

[30] Zack Colman, "Senate extends two biofuels tax credits," *The Hill* (3 Aug. 2012); "Senate Finance Panel Extends Biofuel

Tax Credits," *Energy Overviews* Free Archives website (6 Aug. 2012), available at: http://epoverviews.com/articles/visitor.php?keyword=Cellulosic%20Biofuels

[31] U.S. Department of Energy, *Biomass Multiyear Program Plan.*

[32] Ibid.

[33] "NREL Industry Partners Move Cellulosic Ethanol Technology Forward," National Renewable Energy Lab press release (15 May 2008), available at: http://www.nrel.gov/news/press/2008/600.html

[34] Larry Russo and Neil Rossmeissl, *IBR Platform* (Washington, DC: U.S. DOE, Office of the Biomass Program, EE-2E, 2007).

[35] "NREL Industry Partners Move Cellulosic Forward."

[36] "DuPont Cellulosic Ethanol Biorefinery Project, United States of America," Chemicals-Technology.com website, available at: http://www.chemicals-technology.com/projects/dupont-cellulosic-ethanol-biorefinery-project/

[37] Tennessee Department of Labor and Workforce Development, *Growing Green: The Potential for Green Job Growth in Tennessee* (Nashville, TN: Employment Security Division, Labor Market Information Section, Nov. 2008).

[38] Carly Harrington, "TennEra looks to make maximum use of plants," *The Daily Herald* (Columbia, TN), originally appeared in *The Knoxville News-Sentinel* (29 Sept. 2014).

[39] "DuPont Nevada Site Cellulosic Ethanol Facility," DuPont Biofuel Solutions website, available at: http://biofuels.dupont.com/cellulosic-ethanol/nevada-site-ce-facility/

[40] Note that these numbers do not include models from previous years, which are still circulating in the market.

[41] "Alternative Fuels Data Center," U.S. Department of Energy website, available at: http://www.afdc.energy.gov/data/tab/fuels-infrastructure

[42] Ibid.

[43] "Access to alternative transportation fuel stations varies across the lower 48 states," U.S. Energy Information Administration website (30 Apr. 2012), available at: http://www.eia.gov/todayinenergy/detail.cfm?id=6050

[44] "Compliance Methods for State and Alternative Fuel Provider Fleets," U.S. Department of Energy website, available at: http://www1.eere.energy.gov/vehiclesandfuels/epact/compliance_methods.html

[45] "AFV Acquisitions by Regulated Fleets (by Fuel Type)," U.S. Department of Energy website, Alternative Fuel Data Center, available at: http://www.afdc.energy.gov/data/widgets/10354

[46] "Renewable Fuel Standard (RFS) Program," U.S. Department of Energy website, Alternative Fuels Data Center, available at: http://www.afdc.energy.gov/laws/law/US/390

[47] "Energy Policy Act of 2005," U.S. Department of Energy website, Alternative Fuels Data Center, available at: http://www.afdc.energy.gov/laws/epact_2005

[48] "U.S. Bioenergy Statistics," U.S. Department of Agriculture website, Economic Research Service, available at: http://www.ers.usda.gov/data-products/us-bioenergy-statistics.aspx

[49] "RFS2 EMTS Informational Data," U.S. Environmental Protection Agency website, available at: http://www.epa.gov/otaq/fuels/rfsdata/

[50] Mark Gehlhar, Ashley R. Winston, and Agapi Somwaru, *Effects of Increased Biofuels on the U.S. Economy in 2022* (Washington, DC: U.S. Department of Agriculture, Economic Research report Number 102, Oct. 2010).

[51] U.S. Department of Energy, *Biomass Multiyear Program Plan*, pp. 1-13.

[52] "Florida Project Produces Nation's First Cellulosic Ethanol at Commercial-Scale," U.S. Department of Energy press release (31 July 2013), available at: http://www.energy.gov/articles/florida-project-produces-nation-s-first-cellulosic-ethanol-commercial-scale-0; Laura J. Vimmerstedt and Brian W. Bush, *Effects of Deployment Investment on the Growth of the Biofuels Industry* (Golden, CO: National Renewable Energy Laboratory, Technical Report NREL/TP-6A20-60802, Dec. 2013).

[53] For a more detailed summary of various cellulosic feedstock market price estimates, see Miguel A. Carriquirya, Xiaodong Dub, and Govinda R. Timilsina, "Second generation biofuels: Economics and policies," *Energy Policy* 39, no. 7 (2011): pp. 4222-4234.

[54] Christina Williams, "ZeaChem closes $25M investment, touts Boardman biorefinery," *Portland Business Journal* (29 Oct. 2012).

ABOUT THE AUTHORS

Miles Brundage
Miles Brundage is a Ph.D. student in Human and Social Dimensions of Science and Technology at Arizona State University. His research focuses on the societal implications of artificial intelligence, robotics, and energy technologies. He is currently serving as Research Assistant for the Virtual Institute of Responsible Innovation and Editorial Assistant for the *Journal of Responsible Innovation.*

Travis Doom
Travis Doom earned his undergraduate degree in engineering at Arizona State University. He has spent the last several years with the Consortium for Science, Policy & Outcomes office in Washington, DC working across a wide range of science and technology policy issues.

Jen Fuller
Jen Fuller is a Ph.D. candidate in the Environmental Social Science program at Arizona State University. She currently thinks and writes about the social, political, and cultural dynamics of community renewable energy.

Eric B. Kennedy
Eric B. Kennedy is a Ph.D. student at the Consortium for Science, Policy & Outcomes at Arizona State University. His research focuses on finding improved strategies to foster collaboration in solving complex & interdisciplinary real world problems. A graduate of the inaugural Bachelor of Knowledge Integration cohort at the University of

Waterloo, Kennedy regularly brings together disparate disciplines in his academic work. He maintains active research projects in the fields of innovation systems & social innovation; collaboration across communities, knowledge systems, and sectors; and interdisciplinarity & sociotechnical integration. He is a 2014 Generation Fellow at The Breakthrough Institute and Equinox Fellow with the Waterloo Global Science Initiative.

Frank N. Laird
Frank N. Laird is Associate Professor and Director of the International Studies degree at the Josef Korbel School of International Studies at the University of Denver. His research has ranged widely over energy policy, climate change policy, the interactions between the public and governments over environmental issues, and the history and institutions of science and technology policy. His publications include *Solar Energy, Technology Policy, and Institutional Values* (Cambridge University Press, 2001) and numerous articles in academic and professional journals. His most recent work has analyzed the opportunities and challenges for an energy transition away from fossil fuels and has questioned the simplistic assumptions that underlie many discussions of energy transitions.

Jason O'Leary
Jason O'Leary is a doctoral student in the Human and Social Dimensions of Science and Technology program at Arizona State University and the Consortium for Science, Policy & Outcomes. He is also a graduate scholar at the NSF-DOE Engineering Research Center, Quantum Energy and Sustainable Solar Technologies. He researches social sustainability issues in energy from a complex systems and futures-based perspective.

Heather M. Ross
Heather M. Ross, DNP, is an adult nurse practitioner with a specialized focus in cardiac arrhythmias and implanted

devices that rely on improvements in battery technology to yield improved patient experiences. She is a Ph.D. student in Human and Social Dimensions of Science and Technology at Arizona State University, studying practice and policy surrounding medical technologies. In addition, she is a faculty member at Arizona State University in the College of Nursing and Health Innovation, Center for Sustainable Health at Arizona State University, and Professional Masters in Science and Technology Policy programs.

Daniel Sarewitz
Daniel Sarewitz is co-director of the Consortium for Science, Policy & Outcomes and Professor of Science and Society at Arizona State University. His work focuses on revealing the connections between policy decisions, knowledge creation and innovation, and social outcomes. His most recent book is *The Techno-Human Condition* (MIT Press, 2011, co-authored with Braden Allenby). He is also a regular columnist for *Nature* and the author of *Frontiers of Illusion: Science, Technology and the Politics of Progress.*

Aubrey Wigner
Aubrey Wigner is a Ph.D. student in Human and Social Dimensions of Science and Technology at Arizona State University. His research focuses on integrating makerspaces and hackerspaces with higher education to enhance learning through hands-on interdisciplinary practices. He is also interested in how organizations and individuals engage in technological innovation.

Made in the USA
Middletown, DE
12 November 2026

36932795R10109